土木工程专业研究生系列教材

Technical English in Civil Engineering

土木工程科技英语

夏冬桃　鲁修红　主　编
余泽川　胡　琴　王　歌　刘耀东　副主编

中国建筑工业出版社

图书在版编目(CIP)数据

土木工程科技英语/夏冬桃，鲁修红主编.—北京：中国建筑工业出版社，2019.8
土木工程专业研究生系列教材
ISBN 978-7-112-23945-0

Ⅰ.①土… Ⅱ.①夏…②鲁… Ⅲ.①土木工程-英语-研究生-教材 Ⅳ.①TU

中国版本图书馆CIP数据核字(2019)第131427号

本书基于"兴趣是最好的老师"的理念，内容包括：科技论文写作、纽约环保摩天楼、中国超大桥梁、德国高速公路、京沪高速铁路、英法海底隧道、芝加哥地底城、香港海上机场、抗灾建筑与防震工程、上海中心大厦BIM技术应用、FIDIC（国际咨询工程师联合会）合同条件共11个土木工程前沿专题以及"一带一路"国际合作高峰论坛开幕式主旨演讲报告、实验室安全工作手册等内容。本书内容涵盖土木工程各专业领域：结构体系与性能设计、结构试验与数值模拟、智能建造与防灾减灾、智能交通和智能材料、绿色建筑和BIM技术、学术论文写作以及国际项目合作等方面内容的交叉融合创新，使读者深入了解安全、实用、经济、耐久、节能和环保等土木工程的科学规律，并掌握相关科技知识，深刻理解新理论、新技术、新材料和新工艺与土木工程发展的关系。

本书以任务驱动教学法为编写指导思想，采用新颖的编写方式，每个单元按照引导活动、阅读活动、听力活动、口语活动和写作活动展开编写，读者通过阅读书中内容、词语词组记忆、视频听力活动、课堂讨论以及课后文献阅读和写作活动，加强科技英语"听、说、读、写"综合能力的提高。根据不同课时要求，教师可做适当调整，有选择地使用教材。

本书可作为建筑与土木工程相关专业的研究生以及优秀本科生学习科技英语的教材，也可作为土木工程前沿科技的双语教学的参考教材，还可供从事土木工程设计、施工、管理以及研究的专业人员学习科技英语参考使用。

本书配备教学课件，有需要的读者可通过发送邮件（邮件主题请注明《土木工程科技英语》）至jiangongkejian@163.com索取。

责任编辑：赵 莉 王 跃
责任校对：李欣慰

土木工程专业研究生系列教材
土木工程科技英语
夏冬桃 鲁修红 主 编
余泽川 胡 琴 王 歌 刘耀东 副主编

*

中国建筑工业出版社出版、发行（北京海淀三里河路9号）
各地新华书店、建筑书店经销
北京红光制版公司制版
天津安泰印刷有限公司印刷

*

开本：787×1092毫米 1/16 印张：16¾ 字数：415千字
2019年8月第一版 2019年8月第一次印刷
定价：**60.00**元（赠课件）
ISBN 978-7-112-23945-0
(34228)

版权所有 翻印必究
如有印装质量问题，可寄本社退换
（邮政编码100037）

前言

2018年6月24日，土木工程专业正式纳入我国工程教育专业认证体系及《华盛顿协议》名单，此举将有助于土木工程专业技术人员跨境流动和执业，支撑"一带一路"国家战略的实施。2019年4月25日至27日，第二届"一带一路"国际合作高峰论坛在北京举行，这是一次具有标志性意义的国际盛会。共建"一带一路"为世界各国发展提供了新机遇，也为中国开放发展开辟了新天地。21世纪的新材料、新结构、新技术、新工艺以及新设备日新月异。节能技术、信息控制技术、生态技术等与土木工程结合日益紧密，土木工程正成为许多新技术的复合载体。土木工程所采用的结构体系与性能设计、结构试验与数值模拟、智能建造与防灾减灾、智能交通和智能材料、节能环保以及BIM技术的应用正成为代表一个国家土木工程技术发展水平的重要标志。

不断地提升"听、说、读、写"的英语综合能力将有助于土木工程专业技术人员的职业发展。"授人以鱼，不如授人以渔"。本书编者根据国内外土木工程科技前沿发展现状，精心选择和设计教学单元，有助于提高学生的学习热情，举一反三、熟能生巧，力求使学生逐步掌握土木工程一般词汇和常用术语、结构计算分析、设计施工的一般表达方式，提高英文科技文献的阅读和写作能力以及口语交流能力，对土木工程领域的科学和工程问题进行有效分析和表达，以满足日益增长的国际项目合作与科技交流的时代要求，为海外项目建设和城市发展贡献应有的力量。

本部教材具有三个特色：

1. 前沿性与实用性兼容并举

通过引入：世界上最节约环保的大厦、21世纪之初的中国超大桥和跨海大桥、全球最安全的高速公路、京沪高速铁路、世界第二长的海底隧道以及海底段世界最长的铁路隧道、无限大工程——地底城规划、全球最繁忙的海上机场、防震工程未来科技创想、中国具有里程碑意义的BIM技术项目共9个土木工程前沿专题的情景视听和各项任务驱动学习，以及科技论文写作技巧、实验室安全工作手册、FIDIC（国际咨询工程师联合会）合同条件解读等实用专题，使读者了解土木工程师在公众健康、公共安全、社会文化、法律法规、生态环境以及可持续发展等方面应承担的责任和义务；深入了解安全、实用、经济、耐久、节能和环保等土木工程的科学规律，并掌握相关科技知识；深刻理解新理论、新技术、新材料和新工艺与土木工程发展的关系。知识前沿，信息量大，实用性强。

2. 积极调动读者的融合思维能力，充分发挥语言学习的联想作用

全球最伟大的土木工程项目是如何展开的？每项工程背后的精彩以及挑战性体现在哪些方面？作为设计师、工程师、项目经理以及科技人员，如何克服各种挑战打造现代最高、最长、最快和最复杂的土木工程？大量丰富而生动的工程案例包括纽约环保大厦、中国超大桥梁、德国高速公路、京沪高速铁路、英法海底隧道、芝加哥地底城、香港海上机场、上海中心大厦等全球最伟大的工程，为读者提供体验、实践和感悟问题的情境，将科技知识融入公共英语学习的语境，将设计与施工、试验与分析、机遇与挑战、思考与创新紧密联系，有助于读者不断提升"听、说、读、写"的英语综合应用能力。

3. 专业知识、语言学习与文化修养融于一体

英法海底隧道、芝加哥地底城、京沪高速铁路、抗震建筑——地震工程内容源于美国 Discovery 频道（全球最具影响力的科学探索频道）的科技新知节目：伟大工程、无限大工程、建筑奇观、工程大突破；香港海上机场、纽约环保大厦、中国超大桥梁、德国高速公路源于美国 National Geographic 频道的最新科学与尖端科技节目：工程典范、伟大工程、伟大建筑巡礼等。课文根据原视频节目整理编辑而成，并分成 n 小节，汇编主要的专业词汇表、精彩片段赏析、相关的研究方向等便于读者学习。学生在阅读课文、学习视频与查阅科技文献的同时，将专业英语与西方文化结合起来，在丰富专业语言知识的同时，扩大文化交流和生态保护的视野。

本教材以任务驱动教学法为编写的指导思想，更利于读者专业英语"听、说、读、写"综合能力的提高。教材中每一个单元都可以通过 6 个阶段来展开教学：引导活动→阅读活动→听力活动→口语活动→阅读活动→写作活动。引导活动主要以提问的方式引导学生进入这一单元的主题。听力活动包括：播放小视频、关键词速记、复述视频内容、重复播放精彩小视频等方式。口语活动是听力活动的自然延伸和拓展，要求学生积极参与精彩片段的赏析、专业词汇的理解记忆以及问题讨论和小组展示活动（Presentation），教师做必要的引入和指导。阅读活动和写作活动由教师布置课外任务（作业），学生围绕任务查阅科技文献后撰写科技短文。

感谢湖北工业大学土木建筑与环境学院的资助，感谢国际化示范学院的经费资助。

限于编者的水平，书中的不足之处在所难免，还需要在今后的教学和研究的工作实践中不断加以改进和完善，敬请专家和读者多多批评指正。

编者　湖北工业大学　夏冬桃
2019 年 7 月

CONTENTS 目录

Unit 1 Academic Paper Writing 学术论文写作 1
  1.1 General Description 一般说明 2
  1.2 Title, Author/Affiliation and Keywords
     标题、作者/单位以及关键词 3
  1.3 Abstract: Category and Structure
     摘要：分类和结构 4
  1.4 Abstract: Writing Techniques 摘要的写作技巧 6
  1.5 Introduction 概述 8
  1.6 Texual Development at the Lexical Level of the Main Text
     正文的词汇层面 11
  1.7 Texual Development at the Syntactical Level of the Main Text
     正文的句法层面 13
  1.8 Language Techniques at the Discourse Level of the Main Text
     语篇层面的语言技巧 13
  1.9 Results, Discussion and Conclusion 结果、讨论和结论 14
  1.10 Acknowledgements, Illustrations, Appendice and References
     致谢、图表、附录和参考文献 22

Further Reading
Reading Material 1: Top 10 Avoidable Mistakes as an Author
应避免的 10 条写作错误 22
Reading Material 2: Writing for an Academic Journal: 10 Tips
学术论文写作的 10 条建议 24
Reading Material 3: Civil Engineering Lab Safety
土木工程实验室安全须知 27
Reading Material 4: 学术论文写作 30

Unit 2 Mega Structures: Skyscraper
    伟大工程巡礼：纽约环保摩天楼 35
  Teaching Guidance 36
  2.1 Energy-saving Systems Combined into Skyscraper
     综合多种节能系统的摩天楼 36
  2.2 The Best Way to Get through Rock 最佳岩石开挖方式 37
  2.3 Steel Beam Made from Recycled Material (Scrap Metal)
     钢梁来自再生建材（废金属）............ 38

2.4	Blast Furnace Slag Replaces Regular Cement 高炉矿渣取代普通水泥 … 39
2.5	Utilizing Natural Air & Green Space 自然空气利用和绿色空间 …………… 41
2.6	Roof Cooler, Gray Water of Capturing & Reusing 屋顶降温和废水收集再利用 ……………………………………………… 42
2.7	Allowing Light to Enter While Blocking out Heat 引进光源和隔热全垒打 … 43
2.8	Infinite Imagination of Green Building in the Future 未来绿色建筑的无限畅想 ……………………………………………………… 46

Words and Expressions ……………………………………………………………… 48
Activities——Discussion, Speaking & Writing ……………………………………… 49

Unit 3 Chinese Super Bridges 中国超大桥梁 ……………………………… 53

Teaching Guidance ………………………………………………………………… 54

3.1	The World's Longest Arch Bridge in 2003 2003 年全球最长的拱桥……… 54
3.2	The Challenge of Building the Arch 建造桥拱的巨大挑战 ………………… 55
3.3	The Tied Arch to Prevent Collapse 拉杆拱防止倒塌 ……………………… 57
3.4	The World-class Suspension Bridge 世界级的悬索桥 ……………………… 58
3.5	Big Challenges: Ships, Typhoons & Earthquakes 大挑战：轮船，台风和地震 …………………………………………………… 58
3.6	Key Approach of Fixing the Main Cables 固定主缆索的关键技术 ………… 59
3.7	The System of Main Cable Construction 搭建主缆的系统 ………………… 62
3.8	Key of Wind Resistance Technology 桥梁抗风技术的关键 ………………… 63
3.9	The Longest Sea-crossing Bridge in the World 全球最长的跨海大桥 …… 64

Words and Expressions ……………………………………………………………… 65
Activities——Discussion, Speaking & Writing ……………………………………… 67

Unit 4 German Highway 德国高速公路 ……………………………………… 71

Teaching Guidance ………………………………………………………………… 72

4.1	Unlimited Speed & Precision Engineering 德国的无限速和精准的工程学 ……………………………………………… 72
4.2	The Safest Highway in the World 全球最安全的高速公路 ………………… 72
4.3	Design and Maintenance for Autobahn 德国高速公路的设计和维护 …… 73
4.4	Perfect Monitoring System 完美监控系统 …………………………………… 75
4.5	Efficient Road Rescue Team 高效的道路救援队 …………………………… 77
4.6	Efficient Helicopter Casualty Rescue 便捷直升机伤患救援 ……………… 78
4.7	Autobahn Control Center 高速公路控制中心 ……………………………… 79
4.8	Meticulous Maintenance & Well-designed Highway 精准设计和精心维护 ………………………………………………………… 81
4.9	Top One Driving Technology in the World 全球一流的驾驶技术 ………… 82
4.10	The Unity between Drivers & Cars 人车合一的境界 ……………………… 83

Words and Expressions ……………………………………………………………… 83

Activities——Discussion, Speaking & Writing ········· 84

Unit 5 Beijing-Shanghai High-speed Railway
  京沪高速铁路 ········· 87
  Teaching Guidance ········· 88
  5.1 The Longest High-speed Railway in the World in 2010
       2010 年全球最长的高速铁路 ········· 88
  5.2 Trains Body Design Compotition 车型设计大赛 ········· 88
  5.3 The Strongest Windscreen in the World 全球最强劲的玻璃窗 ········· 89
  5.4 Bogie Ensures Perfect Balance of Train 转向架保证列车完美平衡 ········· 90
  5.5 Railway Bridge to Support the Fastest Trains in the World
       支撑全球最快列车的铁路桥 ········· 92
  5.6 The Slabs for High-speed Track 高速铁路轨道板 ········· 94
  5.7 Speed Sensors & Central Computers 速度传感器和中央控制电脑 ········· 95
  5.8 The Large Roof Ensures Transferring Convenience
       大屋顶保证换乘更便捷 ········· 96
  5.9 Prevent Resonance Phenomenon of Trains from Damaging
       防止列车共振现象的损害 ········· 97
  5.10 Great Engineering Significance 伟大的工程意义 ········· 98
  Words and Expressions ········· 99
  Activities——Discussion, Speaking & Writing ········· 99

Unit 6 The Channel Tunnel 英法海底隧道 ········· 103
  Teaching Guidance ········· 104
  6.1 Engineering Background 工程背景 ········· 104
  6.2 Main Tunnel and Auxiliary Tunnel 主隧道和附属隧道 ········· 105
  6.3 Tunnel Digging 隧道的挖凿工程 ········· 106
  6.4 Transport System 运输体系 ········· 108
  6.5 Investment and Bidding 投资和招标投标 ········· 109
  6.6 Drainage Facilities 排水设施 ········· 109
  6.7 Railway Control Center 铁路控制中心 ········· 112
  6.8 Fire Protection 消防保护措施 ········· 113
  6.9 Maintenance Station 维修站 ········· 114
  6.10 Tunnel Customs 隧道海关 ········· 114
  Words and Expressions ········· 115
  Activities——Discussion, Speaking & Writing ········· 117

Unit 7 Chicago Underground 芝加哥地底城 ········· 121
  Teaching Guidance ········· 122
  7.1 Underground's Development Background 地下城的开发背景 ········· 122

7.2 An Example for Underground under Amsterdam's Canal System
阿姆斯特丹运河系统地下空间示例 …………………………………………… 123
7.3 Excavation of Various Layers of Geology 不良混合地质的开挖技术 …… 125
7.4 Effective Technology for Excavating Stratum 挖掘岩层的高效技术 ……… 126
7.5 Access to the Underground without Disturbing the City Life Above
零干扰城市运行的施工通道 ………………………………………………… 126
7.6 Underground Construction in Moscow 莫斯科的地下建设 ……………… 127
7.7 Ventilation Technology for Underground Engineering 地下工程的
通风技术 ……………………………………………………………………… 129
7.8 Advantages and Disadvantages of Living Underground
地下城的优势和劣势 ………………………………………………………… 130
7.9 Simulating the Day and Night Lighting System 模拟昼夜照明系统 ……… 131
7.10 Concerns to Live Underground 对住在地下城市的关注 ………………… 132
Words and Expressions ………………………………………………………… 134
Activities——Discussion, Speaking & Writing ………………………………… 136

Unit 8 Hong Kong's Ocean Airport 香港海上机场 …………………………… 139
Teaching Guidance ……………………………………………………………… 140
8.1 The Busiest Airport in the World 全球最繁忙的机场 ……………………… 140
8.2 Large-scale Reclamation & Land-making Projects by an Ancient
Invention 大型填海造陆工程 ………………………………………………… 140
8.3 The Greatest Challenge for Maritime Airports 海上机场的最大挑战 …… 141
8.4 The Wind Shear Precaution by Radar and Light
雷达和光学技术预防风切 …………………………………………………… 142
8.5 The Light Roof without Pillars to Support 没有柱子支撑的轻型屋顶 …… 143
8.6 Luggage Handling System 行李处理系统 ………………………………… 146
8.7 Wind Tunnel Tests 风洞试验 ………………………………………………… 147
8.8 Flexible Connection between Roof & Wall 屋顶和墙壁的柔性连接 …… 148
Words and Expressions ………………………………………………………… 149
Activities——Discussion, Speaking & Writing ………………………………… 150

Unit 9 Engineering against Earthquakes 防震工程 …………………………… 153
Teaching Guidance ……………………………………………………………… 154
9.1 Learning from New Events & Disasters 新知识来自新灾难 ……………… 154
9.2 Taiwan Earthquake on September 12 台湾 9.12 大地震 ………………… 156
9.3 Different Earthquake Damage Modes 地震破坏的不同模式 …………… 158
9.4 Earthquake Precautions in California 加利福尼亚的地震预防 …………… 159
9.5 Effective Seismic Measures for General Engineering
一般工程的有效抗震措施 …………………………………………………… 161
9.6 Earthquake Resistance in High-tech Industry Building
高科技工程的尖端抗震技术 ………………………………………………… 162

9.7　Seismic Engineering Science & Technology in Future
　　　未来地震工程的科技畅想 ……………………………………………………… 165
　　Words and Expressions ……………………………………………………………… 167
　　Activities——Discussion, Speaking & Writing ……………………………………… 169

Unit 10　Application of BIM in Shanghai Tower
　　　上海中心大厦 BIM 技术应用 ……………………………………………… 171
　　Teaching Guidance …………………………………………………………………… 172
　　10.1　What is BIM　BIM 简介 ……………………………………………………… 172
　　10.2　BIM throughout the Project Life-cycle　贯穿整个项目生命周期的 BIM … 172
　　10.3　Anticipated Future Potential　BIM 的预期未来潜力 ……………………… 174
　　10.4　Project Introduction of Shanghai Tower　上海中心大厦项目简介 ………… 175
　　10.5　The Challenge of Shanghai Tower Project
　　　　　上海中心大厦项目面临的挑战 ………………………………………………… 175
　　10.6　Overview of BIM Implementation　BIM 实施概述 ………………………… 176
　　10.7　Geometric Design of Shanghai Tower　上海中心大厦结构的几何设计 …… 176
　　10.8　Comprehensive Pipeline Design in 3D Environment
　　　　　三维环境下的综合管道设计 …………………………………………………… 178
　　10.9　Use BIM to Arrange the Construction Schedule
　　　　　使用 BIM 安排施工进度 ……………………………………………………… 179
　　10.10　Use Models to Control Construction Quality　采用模型控制施工质量 … 179
　　Words and Expressions ……………………………………………………………… 180
　　Activities——Discussion, Speaking & Writing ……………………………………… 181

Unit 11　FIDIC Contract in Overseas Project Management
　　　FIDIC 合同在海外项目管理的应用 ……………………………………… 183
　　11.1　Brief Introduction of FIDIC　FIDIC 简介 …………………………………… 184
　　11.2　Standard Forms and Applicability of FIDIC Contract
　　　　　菲迪克合同的标准格式及其适用范围 ………………………………………… 184
　　11.3　Accurate Definition of FIDIC Contract Terminologies
　　　　　菲迪克合同术语的准确释义 …………………………………………………… 185
　　11.4　Documentary Management　文件管理 ……………………………………… 187
　　11.5　Time Bars under FIDIC Contract
　　　　　菲迪克合同条件下的时间节点问题 …………………………………………… 188
　　11.6　Contractor Shall Assume the Primary Liability for Delay
　　　　　工期延误时承包商（施工方）的主要责任 …………………………………… 191
　　Words and Expressions ……………………………………………………………… 196
　　Further Reading ……………………………………………………………………… 199
　　　Reading Material 1: Working Together to Deliver a Brighter Future for Belt and Road
　　　　Cooperation

齐心开创共建"一带一路"美好未来 199
Reading Material 2: CSCEC Works to Expand a Happy Living Environment
中国建筑,拓展幸福空间 200

Appendixes, References & Acknowledgements
附录,参考文献和致谢 203

Appendix 1: Reference Translation for Parts of Paragraphs
附录1:精彩段落参考译文 204

Appendix 2: Common Terms in Civil Engineering
附录2:土木工程常用术语 238

Appendix 3: Network Resource
附录3:网络资源 253

References 参考文献 254

Acknowledgements 致谢 256

Unit 1
Academic Paper Writing

1.1 General Description

1. Basic Types of Academic Papers

Generally speaking, an academic paper is a formal piece of writing in which academics present their views and research findings on a chosen topic, they can be divided into basic subject papers and applied technical papers. In the same academic subjects, they can be divided into basic theory or application papers.

Academic paper contains four types and they are **academic report**, **research paper**, **course paper**, and **thesis/dissertation.**

Whatever a paper may be classified according to different criteria, the task of the writer may, in most cases, be the same: to do research on a particular topic, gather information on it, critically examine the issue (s), and report the finding of the research.

2. Structure of an Academic Paper

A publishable academic paper in English is supposed to include Title, Abstract, Keywords or Indexing Terms, Introduction, Literature Review, Research Methodologies and Procedures, Results and Findings, Discussion, Conclusion, References and Appendixes.

Writing for academic purpose in English concerns how writers meet the requirements of international academic communities. Table 1-1 below are the major parts of an academic paper and their respective functions.

The components of an academic paper and functions   Table 1-1

Items	Main Functions
Title	Summarizing the main text of the paper; Attracting readers
Abstract	Explaining the background, topic, approach, conclusion and significance; Summarizing the paper; Orienting readers to the paper
Keywords or Indexing Terms	Highlighting the focus using noun terms
Introduction	Introducing the research background; Making a research orientation; Stating the research purpose
Literature Review	Reviewing the previous research
Research Methodologies and Procedures	Specifying the methods and procedures used in the study; Explaining the theoretical framework or models and the research design; Clarifying data collection and treatment, experimental apparatus and procedures
Results and Findings	Summarizing the main results and findings; Interpreting or commenting on the most important results with important figures (shown in graphs, tables and diagrams)

continued

Items	Main Functions
Discussion	Expounding the interrelations between the observed facts and their underlying causes; Analyzing the data; Comparing the results with previous studies and the original hypothesis; Developing the hypothesis and speculations; Highlighting the viewpoints; Mentioning the limitations of the study
Conclusion	Summarizing the main points of the study; Presenting the conclusion; Providing implications and suggestions for future work
References	Showing respect for previous work; Facilitating the literature search

Example 1-1

Typical framework of a journal paper drafted as a manuscript.

Title, Authors and affiliations, Keywords, Abstract

1. Introduction, 2. Methods, 3. Results and discussion, 4. Conclusion 5. References

The above items are basic elements of a journal paper. Different journals may require different orders of arrangement and/or subsection names of these elements, whereas the content maintains similar. In other words, the authors should always prepare texts corresponding to each item listed above, regardless of actual format requested by a journal. With all the items ready, the authors would proceed to check journal's requirements and organize those prepared items in the journal's standard format.

1.2　Title, Author/Affiliation and Keywords

The importance of choosing a title is self-evident, the writer should observe the value, the scientific, the innovative and the feasibility principles in choosing a title for an academic paper.

General functions contain "Summarizing the main text, attracting the readers' attention and stimulate their interest, facilitating the retrieval and layout".

Title searches are frequently used by potential readers, to find publications in which they have an interest, therefore, title and key words are crucial.

Linguistic features contain "Using a phrase instead of a sentence, using more nouns, noun phrases and gerunds". Writing requirements contain "be concise, accurate and clear, be brief, be specific, be unified, be standard avoid questions".

Example 1-2

Title: Multicrack detection on semirigidly connected beams utilizing dynamic data.

Keywords:

Connections; Beams; Data analysis; Cracking

1.3 Abstract: Category and Structure

1. Topic Sentence

We may regard the sentence answering the question of "what" in an abstract as the topic sentence. The topic sentence always goes straightforward to the subject or the problem and indicates the primary objectives of the paper.

Useful patterns of topic sentence

The purpose of this paper is...	The primary object of this fundamental research will be to reveal the cause of...
The primary goal of this research is...	
The intention of this paper is to survey...	The main objective of our investigation has been to obtain some knowledge of...
The overall objective of this study is...	
In this paper, we aim at...	With recent research, the writer intends to outline the framework of...
Our goal has been to provide...	
The chief aim of the present work is to investigate the features of...	The writer attempted the set of experiments with a view to demonstrating certain phenomena...
The writers are now initiating some experimental investigation to establish...	The experiment being made by our research group is aimed at obtaining the result of...
The work presented in this paper focuses on several aspects of the following...	Experiments, were made in order to measure the amount of...
The problem we have outlined deals largely with the study of...	The emphasis of this study lies in...

2. Supporting sentences

The topic sentence is usually followed by a few supporting sentences which further specify the subject to be presented. Research methods, experiments, procedures, investigations, calculations, analyses and other significant information will be provided in this part. These supporting sentences, therefore, can be taken as the main body of an abstract.

Useful patterns of supporting sentences

The method used in our study is known as...	The experiment consisted of three steps, which are described in...
	Included in the experiment were...
The technique we applied is referred to as...	We have carried out several sets of experiment to test the validity of...
The procedure they allowed can be briefly described as...	They undertook many experiments to support the hypothesis which...
The approach adopted extensively is called...	Recent experiments in this area suggested that...
Detailed information has been acquired by the writers using...	A number of experiments were performed to check...
The research has recorded valuable data using the newly developed method...	Examples with actual experiment demonstrate...
This is a working theory which is based on the idea that...	This formula is verified by...

3. Concluding sentences

As the ending part of an abstract, concluding sentences usually summarize the research results.

Useful patterns of concluding sentences

In conclusion, we state that...	As a result of our experiments, we concluded that...
In summing up, it may be stated that...	
The results of the experiment indicate that...	This fruitful work gives explanation of...
The studies we have performed showed that...	The writer's pioneering work has contributed to our present understanding of...
The pioneering studies that the writers attempted have indicated that...	The research work has brought about a discovery of...
We carried out several studies which have demonstrated that...	These findings of the research have led the writer to the conclusion that...
The research we have done suggests that...	Our work involving studies of... proves to be effective...
The investigation came out by..., which has revealed that...	The writer has come to the conclusion that...
All our preliminary results throw light on the nature of...	Finally, a summary is given of...
It is concluded that...	

Example 1-3 Abstract

The problem of crack detection has been studied by many researchers, and many methods of approaching the problem have been developed. To quantify the crack extent, most methods follow the model updating approach. This approach treats the crack location and extent as model parameters, which are then identified by minimizing the discrepancy between the modeled and the measured dynamic responses. Most methods following this approach focus on the detection of a single crack or multicracks in situations in which the number of cracks is known. **The main objective of this paper is to address** (**topic sentence**) the crack detection problem in a general situation in which the number of cracks is not known in advance. **The crack detection methodology proposed in this paper consists of two phases** (**supporting sentence**). In the first phase, different classes of models are employed to model the beam with different numbers of cracks, and the Bayesian model class selection method is then employed to identify the most plausible class of models based on the set of measured dynamic data in order to identify the number of cracks on the beam. In the second phase, the posterior (updated) probability density function of the crack locations and the corresponding extents are calculated using the Bayesian statistical framework. As a result, the uncertainties that may have been introduced by measurement noise and modeling error can be explicitly dealt with. **The methodology proposed herein has been verified** (**supporting sentence**) by and demonstrated through a comprehensive series of numerical case studies, in which noisy data were generated by a Bernoulli – Euler beam with semirigid connections. **The results of these studies show that** (**concluding sentence**) the proposed methodology can correctly identify the number of cracks even when the crack extent is small. The effects of measurement noise, modeling error and the complexity of the class of identification model on the crack detection results have also been studied and are discussed in this paper.

1.4 Abstract: Writing Techniques

1. "Five Steps" for Abstract Writing

When a paper is finished, the "Five Steps" in writing an abstract can be used.

Step One: Reading your introduction——to define the context, the topic, and the purpose of the study.

Step Two: Identifying the content of the study——to outline the approach and the procedures of the study.

Step Three: Listing results, findings, conclusion, and significance——to summarize the most important ones and give suggestions of further study.

Step Four: Drafting the abstract——to use your own words or useful ready-made sentence patterns. Eliminate tables, figures, etc. Write in concise English.

Step Five: Checking the draft of the abstract——to study it again with a view to shortening it further to a minimum length. Make sure your abstract is comprehensive representative and readable.

2. "5A Strategy" for Abstract Writing

Unlike the above "Five Steps" for abstract writing, the following "5A Strategy" will be more helpful in writing an abstract when a paper is unfinished.

Before writing your abstract, you need to answer some basic questions.

Q1: What is the general knowledge of your topic in the academic field?
Q2: What aspect of the topic is the paper to focus on?
Q3: What approach do you use to support your main point of view?
Q4: What conclusion will you draw?
Q5: What is the significance of the paper?

A1 (one sentence)
A2 (one topic sentence, and one or two supporting sentences if necessary)
A3 (two or more sentences to give specific information about the approach)
A4 (one sentence, or more if necessary)
A5 (one sentence)

Then the abstract can be completed by using the following formula: Abstract = A1 + A2 + A3 + A4 + A5.

Example 1-4 Abstract

Sample abstract of a paper entitled "Effect of chloride-ion substitution on adhesion of C-S-H layers modeled by molecular dynamics simulations".

(**A1**: **general knowledge of the topic**) Chloride-ion intrusion is regarded as a major threat to mechanical performance and durability of sea-water and sea-sand concrete (SSC). (**A2**: **focus of this paper**) This paper focuses on nanoscale mechanical properties of cement subjected to chloride-ion environment. (**A3**: **methodology**) Based on molecular dynamics technique, this study simulates mechanical performance of double-layer calcium silicate hydrate (C-S-H) with different degrees of chloride-ion substitution. (**A4**: **description of conclusion**) Effect of chloride-ion concentration on adhesion strength of C-S-H would be examined and deteriorating mechanism of chloride ions would be discussed. (**A5**: **significance of this paper**) It is expected that this work could help reveal mechanisms behind the detrimental effect of chloride ions on cement and enrich engineering database for assessment of durability and reliability of SSC.

Example 1-5 Abstract

The problem of crack detection has been studied by many researchers, and many methods of approaching the problem have been developed. To quantify the crack extent, most methods follow the model updating approach. This approach treats the crack location and extent as model parameters, which are then identified by minimizing the discrepancy between the modeled and the measured dynamic responses (**A1**). Most methods following this approach focus on the detection of a single crack or multicracks in situations in which the number of cracks is known. The main objective of this paper is to address the crack detection problem in a general situation in which the number of cracks is not known in advance (**A2**). The crack detection methodology proposed in this paper consists of two phases. In the first phase, different classes of models are employed to model the beam with different numbers of cracks, and the Bayesian model class selection method is then employed to identify the most plausible class of models based on the set of measured dynamic data in order to identify the number of cracks on the beam. In the second phase, the posterior (updated) probability density function of the crack locations and the corresponding extents is calculated using the Bayesian statistical framework (**A3**). As a result, the uncertainties that may have been introduced by measurement noise and modeling error can be explicitly dealt with. The methodology proposed herein has been verified by and demonstrated through a comprehensive series of numerical case studies, in which noisy data were generated by a Bernoulli – Euler beam with semirigid connections. The results of these studies show that the proposed methodology can correctly identify the number of cracks even when the crack extent is small (**A4**). The effects of measurement noise, modeling error, and the complexity of the class of identification model on the crack detection results have also been studied and discussed in this paper (**A5**).

1.5 Introduction

1. General Functions of the Introduction

Every academic paper should have at least one or two introductory paragraphs with or without a particular subtitle. The length or the degree of formality of the paper may decide if the introduction should be a separately-labeled section.

Generally speaking, a useful introduction of a paper should have the following four functions: introducing the background of the subject, showing the research scope, stating the general purpose, explaining the content arrangement.

2. Structural Features of the Introduction

Usually, the structure of an introduction can be described as a funnel-shape (漏斗形), namely starting with the research background, narrowing down to the existing problem, and then focusing on the present research.

Starting with the Research Background

The research background is usually given in the introduction section, and the recent developments in the relevant field are introduced. The way to present this information depends on what the reader already knows.

Introducing the background

Over the past several decades...	Industrial use of... is becoming increasingly common.
The previous work has indicated that...	
Recent experiments by... have suggested...	There have been a few studies highlighting...
Several researchers have theoretically investigated...	
In most studies of..., it has been emphasized with attention being given to....	It is well known that...

Narrowing down to the Existing Problem

After introducing the research background, writers usually make transition to the problem remaining in the previous work to be further studied. Such existing problems may be the ideas that have not yet been mentioned before, the methods that have not been adopted so far, materials that have not yet been discovered in the past, the factors that were previously ignored, and so on.

Presenting existing problems

Great progress has been made in this field. However...	No clear advancement has so far been seen in...
Also... alone cannot explain the observed fact that...	No direct outcome was then reported in...
A part of the explanation could lie in... However...	No such finding could be available in...
The study of... gives rise to two main difficulties: One is... the other is...	So far there is not enough convincing evidence showing...
Despite the recent progress... there is no generally accepted theory concerning...	The data available in literature failed to prove that...
From the above discussion, it appears that at present neither... nor... are known	The theory of... did not explain how many modifications arose...
A major problem... is the harmful effect exerted by...	Although (the research subject)... (the related problem) is yet undetermined
The kind of experiment we have in mind has not been carried out until now.	(The research subject)... However, (the related problem)... remains unsolved
No experiment in this area has suggested that..	(The previous studies) have examined..., (the related problem) is that... despite...
Experiments must be initiated to substantiate...	(Problems in certain research area)... yet (the present solution) has frequently been questioned, because...

Focusing on the Present Research

On the basis of the discussion of the previous research, especially facing the existing problems to be solved, the writer will gradually turn the reader's attention to the present research, by saying primary research objectives, novel ideas, advanced methods, new materials, fresh factors, etc.

Example 1-6 Introduction

Introduction sample of a paper entitled "Effect of chloride-ion substitution on adhesion of C-S-H layers modeled by molecular dynamics simulations".

(**Introduction to background**) In 2018, global production of cement reached billions of tons, requiring aggregate and water resources of the same order of magnitude used as raw materials in concrete mixing. The most common raw materials are sand and fresh water from rivers and lakes. Massive exploitation of fresh water may give rise to a series of environmental problems. Excessive extraction of river sand might lead to the overall down-cutting of river beds, which would cause the levee slope instability during flood season and reduce water supply for residents during dry season. Finding alternatives of traditional concrete raw materials would bring enormous benefits to civil engineering constructions, especially in coastal countries and regions in shortage of fresh water resources.

(**Narrowing down to certain topic**) Use of sea water and sea sand in concrete mixing is of great potential in cost reduction and sustainability in coastal areas, whereas mechanical performance and long-term durability of sea-water and sea-sand concrete (SSC) are of extensive concern to practical engineering applications. To assess and to enhance the viability of using SSC as a major construction material, evaluation and improvement of mechanical performance are essential. A critical factor that casts detrimental effect towards SSC is intrusion of chloride ions, which has been a hot research topic recently. Studies have reported detrimental effects of chloride ions on steel reinforcement and some proposed using non-metal reinforcement to avoid the deterioration of steel. Replacement of steel reinforcement by materials inert to chloride ions is regarded as a promising solution to deal with chloride-ion corrosion. Meanwhile, this solution raises new questions about whether cement phase maintains integral against chloride-ion corrosion. Molecular dynamics method provides a feasible way to investigate response of cement to chloride ions by simulating mechanical behaviors of calcium silicate hydrate (C-S-H), a major hydration product of cement.

(**Focusing on present study**) In the present study, molecular dynamics simulations are performed to investigate nanoscale mechanical properties of cement with different concentrations of chloride ions. A double-layer C-S-H is modeled and adhesion energy between the two layers is calculated. It is noticed that with concentration of chloride ions increases, adhesion energy between C-S-H layers gradually decreases. This could indicate a detrimental effect of chloride ions posted on mechanical properties of C-S-H. Based on the present molecular dynamics study, mechanical performance and viability of SSC in engineering applications is discussed.

Example 1-7 Introduction

The problem of crack detection has been studied by many researchers, and many methods following different approaches and based on different assumptions have been developed. **A comprehensive review of recent developments can be found** (**Introducing the background**) in Sohn et al. (2004). Most of the crack detection methods in the literature have focused on single-crack cases (Cawley and Adams 1979; Rizos et al. 1990; Liang et al. 1991; Narkis 1994; Nandwana and Maiti 1997). For methods that have addressed multicrack situations, it has been assumed that the number of cracks was known in advance. Ostachowicz and Krawczuk (1991) studied the forward problem of a beam structure with two cracks. They expressed the changes in dynamic behavior as a function of crack location and extent. Ruotolo and Surace (1997) studied the inverse problem of the crack detection of beam structures utilizing natural frequencies and mode shapes. They formulated the crack detection process (estimating the location and extent of cracks) as an optimization problem, and solved it by genetic algorithm when the number of cracks was known. Similarly, Law and Lu (2005) proposed the use of measured time-domain responses in the detection of a given number of cracks on a beam structure through optimization algorithms. **The difficulty with this method is that the number of cracks on a beam is generally not known before crack detection(Presenting existing problem).**

Lam et al. (2005) studied the use of spatial wavelet transform in the detection of the crack location and extent of an obstructed beam using the Bayesian probabilistic framework in which there is only one crack on the structural member. **One of the objectives of this paper is to extend the work of Lam et al. (2005) to the multicrack cases in which the number of cracks is not known in advance (Focusing on the Present Research).** The crack detection methodology proposed here is divided into two phases. In the first phase, the Bayesian model class selection method (Beck and Yuen 2004) is employed to identify the number of cracks based on a given set of measured dynamic data. Once the number of cracks has been identified, the posterior probability density function (PDF) of the locations and extents of the cracks are then calculated using the Bayesian statistical framework (Beck and Katafygiotis 1998) in the second phase. Unlike the deterministic approach, which focuses on pinpointing crack locations and extents, the objective of the crack detection methodology proposed in this paper is to calculate the posterior (or updated) PDF of the crack locations and extents. The PDF conveys valuable information to engineers about the confidence level of the crack detection results. The organization of this paper is as follows. The proposed methodology is presented and the related background theories, such as the modeling of the cracked beam, the Bayesian model class selection, and the Bayesian statistical framework, are reviewed. The results of a series of comprehensive numerical case studies, which verify and demonstrate the proposed crack detection methodology, are then reported. The effects of measurement noise, modeling error, and the complexity of the class of identification model on the results of crack detection are then discussed based on the results of these case studies. Conclusions are drawn at the end of the paper.

1.6 Textual Development at the Lexical Level of the Main Text

The main body is very informative and a large part of a paper. The structure of the main body may vary depending on whether it is a theoretical or experimental paper. However, no academic paper is 100 percent theoretical or experimental. In fact, there is no clear demarcation between the two types of academic papers. However, there exist different emphases and distinctive features in papers of the theoretical nature and those of the experimental nature regarding their main body. For example, the typical features of papers of the theoretical nature are embodied in the aspects of description, logical development, etc., while papers of the experimental nature, as the name implies, mainly center on experiments, investigations, and analyses of the results.

Usually, academics may write a paper with a mixture of theoretical and experimental nature. The following table shows an example for the structure of paper of science and technology.

1. Structures of Academic Papers

Paper 1
Opening part without subtitle
An atomic clock to count photons
Progressive pinning-down of photo number
Observing the field-state collapse
Reconstructing photon number statistics
Repeated measurement and field jumps
Beyond energy measurement
Summary

As a matter of fact, in the main body of a paper, especially a paper of an experiment nature, you will usually cover some of the eight items as follows.

(1) Materials used in the experiment

(2) Equipment, devices or instruments

(3) Methods (steps, procedure operation, conditions, etc.)

(4) Calculations

(5) Surveys or investigations (data collection and data analysis)

(6) Results

(7) Discussions

(8) Recommendations

After you have structured your paper, the method of textual development will determine the body of the paper. Thus, it is necessary to master the techniques of using language at the lexical level in this unit, and at a syntactical level and discourse level in 1.7 and 1.8.

2. Language Use at the Lexical Level

As mentioned above, the vocabulary of an academic paper has its own features. They can be specifically summed up as follows: Specialized Terms, Words with Specific Meanings, Greek and Latin Words, Compound Words, Acronyms and Abbreviated Words, Antonyms, Words with New Meanings, Noun Clusters and Nominalization.

Below are more examples of verbs (in the bracket) which are more acceptable in the academic writing: speed up (accelerate), put in (add), breathe in (inhale), think about (consider), take away (remove), increase in amount (accumulate), join together (combine), push away (repel), get together (concentrate), push into (insert).

There are many other words and phrases which have more specific meanings, for example, "suitable for" is more accurate than "all right for". Thus, they prove to be more formal and acceptable in academic papers.

More informal words (formal words) examples are listed below:
without (in the absence of), give (provide), do (achieve),
in the end (eventually), not natural (artificial), make less (decrease),
get bigger (expand), not as good as (inferior to), before this (previously),
all right for/enough for (suitable for), without stopping (unceasingly),
carry (transport), stay alive (survive), do a sum (calculate), act (behave),
about (approximately), upside down (inverted), sometimes (occasionally),
a little (slightly), better than (superior to)...

1.7 Textual Development at the Syntactical Level of the Main Text

The sentence construction of academic papers is comparatively limited in variety and grammatically rigorous. The sentence construction and related elements including tense and mood often used in academic writing feature the following "eight mores".
(1) More Indicative Constructions
(2) More Imperative Construction
(3) More Complex Construction
(4) More "It + be + adj./participle + that clause" Structures
(5) More Attributive Clauses Led by "As"
(6) More Present Simple, Past Simple, and Present Perfect Tenses
(7) More Passive Voices
(8) More Conditional Constructions and Subjunctive Moods

1.8 Language Techniques at the Discourse Level of the Main Text

Coherence is what determines whether your paper makes sense. You must make sure that your sentences and paragraphs are logically presented so as to make yourself understood by your readers. We can compare a paragraph or a paper to a train. The locomotive is the topic sentence in a paragraph or the thesis statement in a paper which gives the train (the paragraph or paper) its direction. Each carriage is a sentence in a paragraph or a paragraph in a paper that must go with the rest of the train, for it must go where the locomotive goes. It is the couplings that hold the carriages together, ensuring that all the carriage will arrive at the same destination as the locomotive. In the same way, you must supply the link between the sentences and paragraphs to give your readers directional signal to indicate what is to come and how it relates to what preceded.

Here are some of the ways of achieving coherence:
(1) By Organizing Arguments Logically
(2) By Using Transitional Words or Phrases Appropriately
(3) By Repeating Key Words or Key Phrases
(4) By Using Synonyms for Key Words or Key Ideas

1.9 Results, Discussion and Conclusion

1. Section of Results

Functions and Contents of Results

The part of results in an academic paper is the summary of your survey, investigation, calculations, experiments, etc. The value of a study lies in the value of its results and the writer's interpretation of them. In this section, the writer should present the essential facts, summarize the significant results, and list the meaningful data.

Besides, the research results are usually presented with corresponding analysis. So in this section the writer should also summarize the essential results and data.

Writing Requirements for Results

The following two points should be kept in mind in writing this section.

Firstly, any data shown in this section must be meaningful. Among all tested variables, only those that affect the conclusion can be included, while those with no such functions should never appear in this section. So sorting out and selecting data should be highly necessary.

Secondly, the presentation of results should be short and clear. This is important because it is the research results that contain new ingredients of knowledge or findings which the writer can claim as his or her own contribution to the academic world. In writing this section, redundancy should be avoided. The following table shows ways for showing the results.

Showing the results

The research we have done suggests an increase in....	Table 5 presents the data provided by the experiments on....
As a result of our experiments, we concluded that....	
This fruitful work gives an explanation of....	This table summarized the data collected during the experiment of...
Our experimental data are briefly summarized as follows...	Some of the writer's findings are listed in tables....
Figure 3 shows the results obtained from studies of....	The direct outcome was then reported in....
Most recent experiments to the same effect have led the writers to believe that....	Sufficient results for... have been observed with the new method...

continued

As Figure 2 shows, the influences of...	As can be seen, the first group of... while the second... The main difference was...
As shown in Table 2, the effect of...	These results suggested that the effect of... was either close to or slightly lower than that of...
Data in Table 1 shows that the influences of...	These changes suggest that the possible reason is...
The effect of... is shown/ summarized in Figure 2	It is considered / found that... these may suggest the reason why...

Example 1-8 Results

Results and discussion section of a paper entitled "Effect of chloride-ion substitution on adhesion of C-S-H layers modeled by molecular dynamics simulations".

(**Show the results**) Results from molecular dynamics simulations show that adhesion energy between double C-S-H layers range from 0.4 J/m^2 to 0.9 J/m^2 as chloride ions vary from 1% to 100%. Dependence of the adhesion energy on concentration of chloride ions is insignificant. (**Compare the results**) The results correspond with existing studies on adhesion energy of C-S-H, which has given a value of around 0.45 J/m^2. The difference between our results and the existing calculations could be ascribed to model difference, method difference and computational error.

(**Analyze the results**) In the plot of adhesion energy against chloride-ion concentration, it is noticed that the adhesion energy scatters randomly as chloride-ion concentration varies. This phenomenon indicates that chloride-ion substitution of hydroxide groups may not affect adhesion strength of C-S-H to a significant level. (**Draw temporary conclusions**) These results imply an insignificant effect of chloride ions on mechanical properties of cement, which might be mixed using sea water and sea sand to cast concrete without mechanical deterioration.

(**Imply limit Actions**) The present work shows that chloride ions in sea water and sea sand have little effect on mechanical properties of cement, according to molecular dynamics simulations of C-S-H double layers with varied degrees of chloride-ion substitution of hydroxide groups. Limitations of this work lie on a lack of considerations of other ions, such as sulfate and magnesium ions of sea water that could affect mechanical properties of cement as well. (**Further study**) Further studies should focus on large-scale simulations and involve considerations of effects from other ions in sea water. It is envisioned that this work could constitute as part of a complete multi-scale studies on sea-water and sea-sand concrete, which would lay the foundation for a comprehensive feasibility study of practical engineering application of SSC.

Example 1-9 Results

Case E tests the proposed methodology in situations in which the crack extent is small (18%

of the overall depth of the beam). Table 7 shows the logarithms of the evidence of M_0 (12,935), M_1(13,135), M_2(13,149), and M_3(13,144) in Case E. It is clear from the table that there are only two cracks on the beam. The optimal parameters and the corresponding COVs are calculated and summarized in the fifth row of Table 9. The identified crack location and extent are again very close to the true values. The proposed crack detection methodology has no problem in identifying the simulated cracks even when the crack depths are small. When the COVs of the identified parameters in Case E are compared with those in Case D, it appears that the uncertainties associated with the crack parameters are relatively higher when the crack extent is small.

2. Section of Discussion

The section of discussion is generally considered to be the most difficult section to write. And many papers are rejected by journal editors because of an imperfect discussion, even though the data of the paper might be both valid and interesting. What is even more likely is that the true meaning of the data in a paper may be obscured by the interpretation presented in the discussion, again resulting in rejection.

Functions and Main Components of Discussion

The section of results functions to summarize the important facts observed by the researcher. The basic functions of discussion are as follows:

(1) Expounding the interrelations among the observed facts;

(2) Showing the underlying causes, the effects, and the theoretical implications of the facts;

(3) Leading to the conclusion.

In writing, the components of discussion usually include:

(1) Analyzing the data: Try to present the relationship between the data. Bear in mind that, in a good discussion, what we need to do is to discuss the results.

(2) Expounding viewpoints: Give the judgment, evaluation and analysis, and show how the results and interpretations agree(or contrast)with previously published works.

(3) Pointing out doubts: Point out any exceptions or any lack of correlation, and state the unsettled points or doubts, limitations, failures, as well as points for attention, if any.

(4) Stating the significance: Point out the theoretical implications of the paper, as well as any possible practical application.

(5) Arriving at a conclusion: Provide the outcome of the research as clearly as possible, and summarize the evidence for the conclusion.

All the above can be regarded as the essential components of the section of discussion.

Writing Requirements for Discussion

According to the functions and basic components stated above, there are some requirements for writing this section.

Firstly, the writer should sufficiently analyze the presented data and point out the relationships. In this respect, the primary purpose of the section of discussion is to show the relationship among observed facts. Faced with the same group of data, different academics may have different conclusions. And it is the point of breakthrough of some scientific research. The writer usually needs to start from a reference to the main purpose or hypothesis of the study, followed by a review of the most important findings, and then show the differences or the matching degree between them, sometimes followed by possible reasons about the differences. So, enough attention should be paid to this point so as to make clear the relationships that can really reflect the research result and can lead to a logical conclusion.

Secondly, the writer certainly has to admit limitations. The limitations of the study may refer to different aspects of the research design. They could be the research methodology, the results of the study, the theoretical models, or the limitation of the samples. It does not mean that the writer simply confesses that he or she has met with failure during the research, but means that he or she states the limitations of the experiment(investigation) conditions, so that the reader will focus on the argument rather than on the fault. To present different aspects of limitation, the writer may need to use different tenses in the writing. For example, the present tense is to show the limitations of the research methodology, model, or data treatment while the past tense is frequently used to present the limitations of what has been done in the experiments.

Thirdly, brief and forceful expressions should be used in the section of discussion to state the conclusion. The conclusions and the inferences should be made based on the results and the study itself. Any overstated conclusion may confuse the readers and thus lessen the persuasiveness of the paper.

Finally, a publishable paper should function as a transition from the previous studies to the further research or practical applications of the results. Thus, it is necessary for a paper to recommend or suggest further study or practical applications in the section of discussion.

Making a comparison between the results and the original hypothesis	
The aim of this research was to propose a novel methodology which... In the present study, it was found that... because...	In this paper, we have reported the significant effect of... The mismatches between the original assumption and the results presented in the study suggest that... Therefore...
This study attempted to investigate whether there are differences in... However, the findings show that...; it is found that... results in...	In this paper, the differences between... were investigated. The results demonstrate that...
This paper has proposed a detailed assessment of... The results presented above show that... This suggests that...	We originally assumed that... The results in this study show that... The reason why... is that...
This study has presented a specific method for measuring...; it is considered that... The results, however, show that...	It was originally assumed that... The differences between... are... This suggests that...
	We originally hypothesized that the effect of... The data in the present study show that... The possible reason may...

Presenting a further explanation of the results

It is possible/likely/unlikely that an erroneous value was attributed to/due to...	This rapid increase / decrease in... is attributed to...
One reason for this could be that inadequate use of... increased...	The enhancement in... may be caused by...
These results can be explained by assuming that the increase/decrease in... resulted in...	It is likely/unlikely that the inaccuracy is attributed to/due to...
This inaccuracy seems to show/indicate that the materials used are...	One reason for this can be explained by assuming that the inadequate use of... increases...

Making conclusions or inferences

These results indicate / suggest / show/imply that...	Our conclusion is that...
	Our data provides evidence that...
The data reported here imply / suggest / indicate / confirm that...	It suggests that...

Implying the limitations of the present study

The number of the participants in this survey was relatively small	The findings may be valid if the above discussed conditions are changed within the accuracy limits
Only three groups of samples were tested in the current study	An experiment employing different approaches might produce different results
Other elements which may cause this change were assumed as the constant in the formula	We recognize that the method adopted in the current study does not cover... due to...
Tests on this parameter with other kinds of participants might yield different results	We readily admit that single short test on this parameter may not fully identify...

Suggesting the practical applications or further study

A further experiment should be conducted with... (a new research method) in order to generalize the effect of.. (the results in the current study)	The results presented in this paper should be useful in.. (a practical area) such as...
Future research could explore the possibility of applying (a new aspect of the theory) to...	Further studies should focus on the practical use of... (the results) in.. (a practical area).
In the future, the effect of... (the unsolved problems of this study) will be examined	We suggest that a series of similar studies be conducted with... (other research methods)
An other interesting topic would be to examine how... (the other aspect of the present study)	We recommend that these experiments be replicated using a wider range of... (different materials or procedures)
An important direction for further work might be to study... (the unsolved question in the study)	In the future, we will investigate the effect of... (the results in the present study) in a series of studies
..could be assessed in studies using other types of... (research materials or procedures)	Researchers of this paper are now conducting experiments with... (other research methods)

Example 1-10 Discussion

This study proposes a new ballast damage detection methodology following the Bayesian approach utilising multiple sets of measured modal parameters from an in-situ sleeper. Using the Bayesian model class selection method, if the selected model class shows a uniform ballast stiffness distribution under the sleeper, the system is treated as undamaged. Otherwise, the ballast at certain Region(s) under the sleeper may be damaged, and the Bayesian model class selection method helps "identify" an "appropriate" class of models to capture the dynamic behavior of the system based on a given set of measurements. The posterior PDF of ballast stiffness at different regions under the sleeper can be calculated by the Bayesian model updating method. The ballast damage can then be detected by comparing the means and standard derivations of ballast stiffness with the reference line. In this study, the reference line is obtained via Bayesian model class selection and model updating using a set of measurements from the undamaged system. Most structural damage detection methods require a reference system. However, it is well known that the set of measurements corresponding with an undamaged (reference) system is usually difficult to obtain. According to the results of the full-scale experimental case study, the ballast stiffness distribution in an undamaged case was uniform under the sleeper, and the ballast stiffness value was very close to the nominal value. **As a result, it is proposed to carry out comprehensive impact hammer tests, and form a database of measured modal parameters** (including updated ballast stiffness) for different types of railway track system under different configurations. The ballast stiffness values in the database can then be treated as the "reference line" in detecting damage to the ballast in **real applications of the proposed methodology (Suggesting the practical applications or further study)**.

One outstanding advantage of the Bayesian approach is the calculation of the marginal posterior PDFs of uncertain model parameters, which provides measures of the uncertainties associated with the model updating (and damage detection) results. Another contribution of this study is the design and build of an indoor test panel. The train service in Hong Kong is extremely busy, and it is almost impossible to carry out vibration test on real ballasted track systems. The indoor test panel allows the simulation of different ballast damage scenarios and vibration measurement can be carried out without time limit.

Several difficulties must be overcome before the proposed methodology can be put into real-world practice (Implying the limitations of the present study). The first involves the modelling of the rail – sleeper – ballast system. The ballast stiffness variation along the concrete sleeper would be continuous in a real-world situation, and the discontinuous "jumps" in ballast stiffness at the interfaces between two regions in the case studies were "artificial". Further research must be carried out for developing a continuous model to describe the distribution of ballast stiffness along a sleeper. In the proposed methodology, it is assumed that the sleeper is undamaged. However, both the sleeper and ballast may be damaged in a real-world situation. Damage to the sleeper would alter the modal parameters of the system and certainly increase the difficulty involved in detecting damage to the ballast. In the experimental case study, modes 1 and 2 were shown to be sensitive to ballast damage, and modes 3 and 4 were sensitive to sleeper damage. By following this direction, the current Bayesian ballast damage detection methodology could be extended to consider possible damage at both ballast and sleeper. Putting the test panel indoors meant that environmental factors such as temperature and humidity could be kept at a consistent level in all of the vibration tests. This is very important in the development stage of a ballast damage detection method. However, the effects of temperature and humidity on the modal parameters of the in-situ sleeper must be studied in detail. This is because changes in temperature and humanity will induce changes in modal parameters of the in-situ sleeper.

3. Section of Conclusion

As the end result of the whole paper, the section of conclusion (also called summary, concluding remarks, etc.) is the final viewpoints drawn by the writer after investigations, experiments inferences, discussions, and so on.

Functions and Main Elements of Conclusion

(1) Summing up: Summarizing the chief facts, data, and other important information in the paper.

(2) Presenting statement of conclusion: Affirming the thesis statement.

(3) Providing statement of recommendation: Recommending or suggesting a further research study or practical use.

Writing Requirements for Conclusion

As the ending part of an academic paper, the conclusion aims to answer the following questions:

(1) Does the conclusion of the present study accord with the original research design? If not, why?

(2) According to the results and discussion, what conclusion or inference may be made? And why?

(3) Does the conclusion accord with those of other researchers? If not, why?

(4) Is there any suggestion for further study?

(5) Are there any practical applications of the results? What are they?

To answer these questions, the writer should follow the points below in writing the conclusion.

If you make a summary, make certain that you actually summarize, with due emphasis, the principal information in the main body of the paper.

Conclusions are convictions based on evidence. If you state conclusions, make certain that: They follow logically from the data presented in the main body of the paper; and they echo with your thesis statement in the introduction or they are exact answer to your research question in the introduction.

If you include recommendations, be sure that they follow logically from the data and conclusions presented earlier, and they should never clash with what you may have expected to do in your introduction.

Graceful termination is achieved when all the materials of the conclusion are smoothly woven together. Be on guard against duplicating large portions of the introduction in the conclusion. Verbatim (一字不差的) repetition can be boring and even create a false unity.

Drawing conclusions

What is more important is that the conclusion provides the answer to the research question posed in the introduction. We can identify the conclusion easily by the sentences below:

(1) On the basis of..., the following conclusion can be made...

(2) From..., we now conclude...

(3) To sum up, we have revealed...

(4) We have demonstrated in this paper...

(5) The results of the experiment indicate...

(6) In conclusion, the results show...

(7) The research work has brought about a discovery of...

(8) Finally, a summary is given of...

(9) These findings of the research have led the writer to the conclusion that...

(10) The research has resulted in a solution of...

Example 1-11 Conclusion

Conclusion section of a paper entitled "Effect of chloride-ion substitution on adhesion of C-S-H layers modeled by molecular dynamics simulations".

From molecular dynamics simulations, several concluding remarks are listed. (**Brief recap of the present study**) In model construction, it is noted that chloride ions may substitute hydroxide groups distributed in interlayer water. Based on this observation, atomistic models of double-layer C-S-H with different levels of chloride-ion substitution are constructed. Following molecular dynamics simulations show that adhesion strength between the two C-S-H layers decreases with increasing levels of chloride ion substitution. (**Conclusion to claim**) The results indicate that adsorption of chloride ions in C-S-H leads to deteriorations of cement in terms of mechanical performance. (**Compare to other studies**) This finding corresponds to existing experimental studies on mechanical properties of hardened cement paste and reveals microscale mechanism of deterioration of cement. (**Further study**) The results also base a foundation for further studies on multi-scale mechanical response of cement to chloride ions by performing coarse-grained simulations at the mesoscopic length scale.

Example 1-12 Conclusion

This paper addresses the problem of crack detection in beams utilizing a set of measured dynamic data. Unlike other crack detection methods in the literature, the proposed methodology is applicable to multicrack cases even when the number of cracks is not known in advance. The proposed methodology relies on the Bayesian model class selection method to identify the number of cracks based on the set of dynamic measurements. The updated PDF of the crack location, extent, and other uncertain model parameters, such as the rotational stiffness for modeling the semirigid behavior of the beam end connections, and the damping ratio, is calculated by the Bayesian statistical framework.

1.10 Acknowledgements, Illustrations, Appendices and References

It is mainly used to extend the indebtedness of the writer to the support or concern from any individual(s) or institution(s) for offering any useful material, specimens, technical know-how, fund, suggestions, or any other kind of enlightenment, etc.

No matter where it is located, the section of acknowledgments always fulfills the function of expressing the appreciation and gratitude of the writer. It is generally the last section of the body of a paper. In some cases, it can be put down as a footnote for the title or the writer's name. As for a thesis or dissertation for degree, it is usually paced prior to the body of the paper on a separate page.

Example 1-13 Acknowledgment

The work described in this paper was fully supported by a grant from the Research Grants Council of the Hong Kong Special Administrative Region, China (Project No. CityU 1190/04E).

Examples in the Chapter refer to the following two papers:

[1] Lam H F, Ng C T and Leung A Y T. Multicrack detection on semirigidly connected beams utilizing dynamic data. Journal of Engineering Mechanics ASCE 2008(134):90-99.

[2] Lam H F, Hu Q and Wong M T. The Bayesian methodology for the detection of railway ballast damage under a concrete sleeper. Engineering Structures 2014(81): 289-301.

Further Reading

Reading Material 1

Top 10 Avoidable Mistakes as an Author

1. Ignoring the journal's instructions authors

(1) Follow all journal instructions regarding word count, page margins, page numbering, spacing, in-text citations, references, abstract, manuscript format, etc.

(2) Ensure that your queries don't indicate that you have not read/followed the journal's instructions for authors.

2. Presenting inconsistent data

(1) Ensure that the numbers and units of measurement presented in the text of your paper are consistent with those presented in tables and figures.

(2) Before submitting your paper to a journal, always check whether the data cited in various sections are an exact match.

3. Ignoring a journal's reference citation policy

(1) Ensure that no references are missing & that none of the listed references are either inaccurate or in correctly formatted.

(2) Assign numbers to references at the citation point, number your reference list and retain unique numbers for references, irrespective of their frequency of use.

4. Revealing participant identity

(1) If your study contains any information that might reveal the identity of the study's participants, ensure that you obtain informed consent from the participants.

(2) If you cannot track a participant or are unable to get consent, then consult the journal editor about the participant details that need to be omitted.

5. Presenting exaggerated conclusions

(1) Your conclusions should be entirely supported by data. Don't make false or overly grand claims about your study's potential to bring about change.

(2) Don't be selective while interpreting results. ; avoid focusing only on positive results while ignoring the negative and neutral ones.

6. Making punctuation and style errors

(1) Try to avoid making common punctuation mistakes. Learn correct punctuation usage such as when to use em dashes, commas, and brackets.

(2) Ensure that you have expanded allabbreviations, avoid using different abbreviation for the same term throughout your paper.

7. Leaving footnotes unexplained

(1) Don't neglect explaining things that seem obvious to you; they might not necessarily seem as obvious as readers.

(2) Explain such details in a footnote, indicating, for example, that the data values presented in the tables are rounded off values.

8. Submitting incomplete or incorrectly filled forms

(1) Ensure that you have correctly and completely filled all the forms as required in the jour-

nal such as those regarding authorship, conflicts of interest, etc.

(2) Ensure that none of the forms are missing; you could avoid unnecessary delays in the publication process by following these guidelines.

9. Engaging in duplicate submission

(1) Don't submit the same manuscript, or even part(s) of it, to 2 or more journals simultaneously. It is regarded as a breach of publication ethics.

(2) This holds true even if the journals involved differ as local and international or if they have different target audiences.

10. Not understanding the copyeditor's changes

(1) Avoid questioning changes made by the journal's copyeditor without first understanding the rationale behind these changes.

(2) Ensure that you read the edited manuscript as well as the copyeditor's cover letter sent along with it, before hastily questioning any changes made.

Reading Material 2

Writing for an Academic Journal: 10 Tips

1. Have a strategy, make a plan

Why do you want to write for journals? What is your purpose? Are you writing for research assessment? Or to make a difference? Are you writing to have an impact factor or to have an impact? Do you want to develop a profile in a specific area? Will this determine which journals you write for? Have you taken their impact factors into account?

Have you researched other researchers in your field – where have they published recently? Which group or conversation can you see yourself joining? Some people write the paper first and then look for a "home" for it, but since everything in your article – content, focus, structure, style – will be shaped for a specific journal, save yourself time by deciding on your target journal and work out how to write in a way that suits that journal.

Having a writing strategy means making sure you have both external drivers – such as scoring points in research assessment or climbing the promotion ladder – and internal drivers – which means working out why writing for academic journals matters to you. This will help you maintain the motivation you'll need to write and publish over the long term. Since the time between submission and publication can be up to two years (though in some fields it's much less) you need to be clear about your motivation.

2. Analyse writing in journals in your field

Take a couple of journals in your field that you will target now or soon. Scan all the abstracts over the past few issues. Analyse them: look closely at all first and last sentences. The first sentence (usually) gives the rationale for the research, and the last asserts a "contribution to knowledge". But the word "contribution" may not be there-it's associated with the doctorate. So which words are used? What constitutes new knowledge in this journal at this time? How can you construct a similar form of contribution from the work you did? What two sentences will you write to start and end your abstract for that journal?

Scan other sections of the articles: how are they structured? What are the components of the argument? Highlight all the topic sentences – the first sentences of every paragraph – to show the stages in the argument. Can you see an emerging taxonomy of writing genres in this journal? Can you define the different types of paper, different structures and decide which one will work best in your paper? Select two types of paper: one that's the type of paper you can use as a model for yours, and one that you can cite in your paper, thereby joining the research conversation that is ongoing in that journal.

3. Do an outline and just write

Which type of writer are you: do you always do an outline before you write, or do you just dive in and start writing? Or do you do a bit of both? Both outlining and just writing are useful, and it is therefore a good idea to use both. However, make your outline very detailed: outline the main sections and calibrate these with your target journal.

What types of headings are normally used there? How long are the sections usually? Set word limits for your sections, sub – sections and, if need be, for sub – sub – sections. This involves deciding about content that you want to include, so it may take time, and feedback would help at this stage.

When you sit down to write, what exactly are you doing: using writing to develop your ideas or writing to document your work? Are you using your outline as an agenda for writing sections of your article? Define your writing task by thinking about verbs – they define purpose: to summarise, overview, critique, define, introduce, conclude, etc.

4. Get feedback from start to finish

Even at the earliest stages, discuss your idea for a paper with four or five people, get feedback on your draft abstract. It will only take them a couple of minutes to read it and respond. Do multiple revisions before you submit your article to the journal.

5. Set specific writing goals and sub – goals

Making your writing goals specific means defining the content, verb and word length for the section. This means not having a writing goal like, "I plan to have this article written by the end of the year" but "My next writing goal is to summarize and critique twelve articles for the literature review section in 800 words on Tuesday between 9am and 10:30am". Some people see this as too mechanical for academic writing, but it is a way of forcing yourself to make decisions about content, sequence and proportion for your article.

6. Write with others

While most people see writing as a solitary activity, communal writing – writing with others who are writing – can help to develop confidence, fluency and focus. It can help you develop the discipline of regular writing. Doing your academic writing in groups or at writing retreats are ways of working on your own writing, but – if you unplug from email, internet and all other devices – also developing the concentration needed for regular, high-level academic writing.

At some point – ideally at regular intervals – you can get a lot more done if you just focus on writing. If this seems like common sense, it isn't common practice. Most people do several things at once, but this won't always work for regular journal article writing. At some point, it pays to privilege writing over all other tasks, for a defined period, such as 90 minutes, which is long enough to get something done on your paper, but not so long that it's impossible to find the time.

7. Do a warm up before you write

While you are deciding what you want to write about, an initial warm up that works is to write for five minutes, in sentences, in answer to the question: "What writing for publication have you done (or the closest thing to it), and what do you want to do in the long, medium and short term?"

Once you have started writing your article, use a variation on this question as a warm up – what writing for this project have you done, and what do you want to do in the long, medium and short term? Top tip: end each session of writing with a "writing instruction" for yourself to use in your next session, for example, "on Monday from 9 to 10am, I will draft the conclusion section in 500 words".

As discussed, if there are no numbers, there are no goals. Goals that work need to be specific, and you need to monitor the extent to which you achieve them. This is how you learn to set realistic targets.

8. Analyse reviewers' feedback on your submission

What exactly are they asking you to do? Work out whether they want you to add or cut something. How much? Where? Write out a list of revision actions. When you resubmit your article, include this in your report to the journal, specifying how you have responded to the reviewers' feedback. If your article was rejected, it is still useful to analyse feedback, work out why and revise it for somewhere else.

Most feedback will help you improve your paper and, perhaps, your journal article writing, but sometimes it may seem overheated, personalised or even vindictive. Some of it may even seem unprofessional. Discuss reviewers' feedback – see what others think of it. You may find that other people – even eminent researchers – still get rejections and negative reviews; any non – rejection is a cause for celebration. Revise and resubmit as soon as you can.

9. Be persistent, thick – skinned and resilient

These are qualities that you may develop over time – or you may already have them. It may be easier to develop them in discussion with others who are writing for journals.

10. Take care of yourself

Writing for academic journals is highly competitive. It can be extremely stressful. Even making time to write can be stressful. And there are health risks in sitting for long periods, so try not to sit writing for more than an hour at a time. Finally, be sure to celebrate thoroughly when your article is accepted. Remind yourself that writing for academic journals is what you want to do – that your writing will make a difference in some way.

Reading Material 3

Civil Engineering Lab Safety

The following reading material shows examples for lab safety in university labs. When conducting experiments, supervisors and students should be highly aware of the safety operation.

1. General Description

Lab for civil engineering research generally features heavy machines driven by electric engines with a high voltage or fuel engines. Involved mechanical forces are often at the magnitude of tens of tons, which(is designed to)break materials easily and far exceed carrying capabilities of normal people. Consequently, any accident would damage equipment, cause injuries or even death. To ensure lab safety, precautions should be observed and taken by all users of laboratory facilities. Here is a list of typical requirements on general behaviors(referred to laboratory manual from department of architecture and civil engineering, City University of Hong Kong).

(1) No smoking.

(2) No running, no playing games, no throwing objects. Disciplinary actions will be taken against offenders.

(3) Safety precautions must always be observed.

(4) Observe any instructions or advice given by the laboratory staff.

(5) Do not attempt to install, correct or operate any apparatus before reading the instructions, seek assistance from laboratory staff in the case of any doubt.

(6) Treat every piece of apparatus with caution.

(7) Beware of others working in the vicinity.

(8) Understand all safety precautions before working in the laboratories.

(9) If in doubt, ask for assistance.

Besides, issues about mechanical and electrical safety, as well as emergency procedures, are discussed in following sections.

2. Mechanical Safety

Users of laboratory facilities are responsible to keep machines in proper working status and to maintain a safe working environment. To achieve this, people working in lab must know how to operate a machine before using it, be clear about working conditions of the machine and have a basic understanding of types of mechanical load that the machine may cause in accidental situations. A critically important rule is not to operate a machine that you are not familiar with. Mechanical equipment, especially heavy ones, may result unexpected movements and forces which lead to dangerous situations. Therefore, it is essential for all users of mechanical equipment to gain a basic understanding of what kinds of movements and forces can be generated by the working machine. Generally, there are several types of forces, including static force, pressure, dynamic force, vibration, cutting forces and torque, The following names a few.

(1) Static force: If the support to this force is not strong enough, it might collapse suddenly.

(2) Pressure: If the safety valve for a pressure vessel, e.g., a storage tank of an air compressor, is not working, the pressure will gradually develop to a point where a sudden explosion is triggered.

(3) Dynamic force: The magnitude of a dynamic force is constantly changing due to acceleration and deceleration of certain parts of a machine, which is strictly untouchable at work.

(4) Vibration: Mechanical vibration can be found when using a machine that is specifically designed to simulate earthquake. Vibration is often hard to control and if there is unusual vibration or noise caused by mechanical movement, you should immediately stop the machine and report to technician.

(5) Cutting forces: Cutting a steel bar is normal for civil engineering experiment. Operator of cutting machines must be clear about direction of cutting forces and make sure that the steel bar being cut is properly fixed. Incorrect operation of cutting machines will cause abnormal behaviors of the machine and unexpected forces which might damage the operator or surroundings.

(6) Torque: Rotation is a common way to generate torque. Rotating wheels turn out to be very dangerous if they are not properly fixed and they might fly out in various directions.

3. Electrical Safety

Some electrical faults, including overload, short circuit and earth faults, may cause electricity hazards. Overload refers to the situation where applied load is greater than design value of the circuit. Short circuit is due to cable fault or external damage to the wiring system. Earth faults refer to short circuit of low impedance between phase and earth system or protective layer. Resulted electrical hazards are listed here.

(1) Electric shock is physical stimulation that occurs when electric current passes through human body. Effect of electric shock depends on magnitude of the current, body parts through which the current flows, duration. The minimum value of current which can be sensed by human body, also known as the threshold of perception, is usually assumed to be 0.5 mA. The "let-go current" threshold, which will result in contractions of muscles and finally lost of voluntary control, is typically 15 mA at 50Hz AC. Current exceeding the "let-go current" flow through chest, head or nerve centers controlling respiration may produce respiratory inhibition. Even stronger current will disturb the normal rhythmic expansion and contraction of heart muscles. Very high current will cause severe burning.

(2) Indirect injuries may also be caused due to involuntary muscular movement, e. g., falling from a ladder due to loss of balance and thrown away by powerful muscular contractions.

(3) Fires of electrical origin is also a potential concern for civil engineering labs. Fires may be caused by overheating of conductors and/or adjacent flammable materials, mainly due to short circuit or current leaking.

To protect our body from electric shock, we can prevent or limit the shock by avoiding dangerous potential difference from being applied across body and by increasing resistance of possible current path. Switching off circuit and isolation is also reasonable approach if someone else is involved in electrical danger.

4. Emergency

An emergency is an abnormal and dangerous situation needing prompt action to control, correct and return to safety. It may be considered as a situation where injuries or damage to property have or could be incurred. In the case of emergency there is no point in waiting and risk panic, whereas a key action is to follow emergency plan. Planning for emergencies aims to ensure safety of occupants of buildings in the event of an emergency and to minimize damage. Goals of an emergency plan are to decrease level of risk to life, property and environment, as well as to control any incident and minimize its effects. Here, discussions on fire event, disaster and electrical injuries are presented.

(1) In the event of fire, we have several rules to observe, including:

1) Remain calm.

2) Alert persons nearby by shouting fire.

3) Evacuate threatened area immediately.

4) Break fire alarm.

5) Attack on fire if it is safe to do so.

6) Close doors to isolate fire and smoke.

7) Exit building while pay attention not to use lifts and assist people in need.

8) Do not re-enter a building on fire until clearance is given.

(2) In the event of disaster, such as leakage of gases, explosion, etc.:

1) Remain calm.

2) Alert people nearby.

3) Evacuate threatened area immediately.

4) Summon laboratory technicians or officers, who are responsible in giving relevant details including nature of the disaster, exact location and people at risk.

5) Do not re-enter threatened area until clearance is given.

(3) In the event of electrical injuries:

1) Do not touch the casualty's flesh with your hands.

2) Break contact (e.g., stand on a wooden box or a rubber/plastic plate and use a broom on wooden chair to push the casualty's body away from electric source).

3) Commence cardiopulmonary resuscitation if you have received relevant training.

4) Alert people nearby and call technicians for help.

Reading Material 4

学 术 论 文 写 作

1. 概述

在国际性学术刊物上发表文章的质量和数量是衡量一个国家、一个机构或专业人员学术水平和学术地位的重要标志。学术论文包括专业报告、研究论文、课程论文、学位论文等。

学术论文是专业人员的研究结果和学术总结。写作一篇学术论文大体上要经历：寻找有价值的思想、确定话题、展开研究、明确写作对象和目的、讨论并完成初稿以及修改与润色等过程。在具体写作之前，有必要先见"林"后见"树"，弄清楚英语学术论文的主要结构和语体特征。

学术论文采用学术语体，它是专业人员用来记录学术积累，展示科研成果，使之得到社会承认并广为传播的有效手段。这种以学术交流为目的的语言是"一种限定的语言，服务于限定领域的经验和行为，这种语言有其自身的语法和词汇特征"。因此，其特点是多用正式语体，包括正式的词汇和短语，客观描述的句型和人称及逻辑严密的语篇结构等。这些语言特点构成了学术语体的基本特征。

学术语体是说理性语言，是严谨推理和复杂思维活动的文字体现。从宏观上把握学术语体的语言风格，熟悉学术语体在遣词、造句语篇等方面的基本特点，对于提高学术论文的写作质量和提高国际发表的成功率，无疑具有重要的意义。

2. 标题、作者/单位、关键词

论文的标题旨在概括全文、吸引读者、便于检索。英文标题一般不使用完整的句子,多用名词、名词性词组成动名词。英文标题常以短语为主要形式,以名词性短语为常见。之所以标题一般不用陈述句,是因为标题主要起标示作用,用陈述句容易使标题具有判断式的语义;此外陈述句不够精练,重点也不易突出。在写作上,标题不应过长,力求简明;避免空洞和笼统,一般少用"提问式"。标题应避免名词与动名词混杂使用,避免硬用非标准化的缩略词语等。总的原则是,标题应确切、简练、醒目。

作者署名,表明文责自负,既便于检索、联系,同时也可以提高作者的知名度。署名采用汉语拼音。署名和单位的格式应统一遵照拟投稿的学术期刊的体例要求。

关键词既有提示文章主题的功能,又有便于检索之作用。关键词多用名词,字数有一定限制。一般来说,关键词主要是从论文的标题和摘要中选择,其书写要规范,符合拟投稿的学术期刊的体例要求。

3. 摘要

论文摘要是论文的缩影,是论文的高度概括和浓缩,旨在使读者迅速了解全文梗概。论文摘要在一定程度上可决定其作者能否获得发表的机会,决定论文能否被期刊选中发表,决定研究能否获得支持,论文摘要能使有关决策人物在没有时间阅读全文的情况下迅速了解论文内容,评估其价值和意义,能帮助作者得到认可、支持、资助,便于开展下一步工作。总之,从论文摘要的基本功能上足见论文摘要的重要性。

论文摘要通常包括描述型摘要、信息型摘要、信息型和描述型相结合的摘要。论文摘要主要包括五个方面的基本内容:(1)研究的背景;(2)研究的目的;(3)研究的方法;(4)研究的结论;(5)研究的意义。

论文摘要在结构上比较固定,大体上遵循"三段式"章法,即通常由开头、展开和结尾三部分构成。开头通常称为主题句。主题句开门见山地点明主题,指出论文的主要内容是什么。展开句进一步阐明主题句的具体内容,指出研究方法、分析过程及论证的要点等,介绍具体过程。结尾句提出论文得出的结果和结论。

论文摘要的写作有两种主要方法,一种是先写摘要,在摘要的基础上撰写论文;一种是后写摘要,依据全文概括归纳出摘要,可分五步进行。写作摘要需要回答这样五个问题:(1)本研究项目在特定研究领域里的学术背景如何(即有哪些相关研究)?(2)本研究的重点是什么?(3)研究方法和技术手段是什么?(4)本研究获得什么样的结果?得出何种结论?(5)本项研究的主要贡献或研究的意义是什么?

论文摘要是多种信息的综合,写作摘要应注意篇幅、语言简明、内容前后一致、时态恰当以及人称、语态得当等写作要求。论文摘要中常见的错误包括:口语化,内容不完整,语言表达重复、单调以及相关信息不均衡等。

4. 论文引言

引言部分的基本功能包括:
(1)论文主题内容,指出论文做了什么研究。主题内容是什么,即 what 的问题。

(2) 限定研讨范围,指出论文主题的界限,参数研究的重点等,即 where 的问题。
(3) 说明论文目的,指出撰写该论文的重要意义,即 why 的问题。
(4) 交代写作过程,指出论文的写作顺序和总体规划,即 how 的问题。

引言部分的结构特点包括：

(1) 从简述研究的背景入手。在引入主题之后,作者往往要列述在特定研究领域里人们已经做了哪些相关的工作,简要综述有关文献,使读者了解本文拟展开的研究与过去所做研究之间的关系,即要回答 What has been done? 的问题。

(2) 以解决现存的问题承转。接下来的引言写作,主要是在以往研究的基础上,指出已有研究中存在的不足、空缺、弱项等问题以及值得改进、尚需更新的地方,即要回答 What has not been done? 的问题。

(3) 以当下研究的问题为焦点。由上述以往研究中存在的问题转而提出本文的研究。这里需要回答的是 What am I (are we) going to do? 的问题。

5. 论文正文

论文正文在句子结构方面的特点可以概括为"八多",即(1)陈述句多；(2)祈使句多；(3)复合句多；(4)用"It"作形式主语的句式多；(5)"As"引出的定语从句多；(6)三种时态(一般现在时、一般过去时和现在完成时)多；(7)被动语态多；(8)条件句及虚拟语气多。

简洁的英文表达是学术论文作者致力于追求的目标之一。使论文文字简洁的方法有很多,主要手段包括：利用非谓语动词,变定语后置为前置,运用名词、动词、形容词、副词和介词短语,利用省略,利用专业词汇,删除冗余表达等。

学术论文的正文部分,主要侧重于叙述实验的原理或运用理论进行分析,叙述研究的各个方面及其关系,交代前因后果、变化等等。为此,该部分的语言侧重于用简洁的文字把事物的性质、特征、成因、关系、功能等解说清楚。

论文正文写作的基础是词汇。英语学术论文的词汇具有专业性强、词义具体、希腊语和拉丁语词语多、多用复合词、缩略词数量大、反义词广泛应用、旧词新义普遍以及名词化等特征。论文正文写作在定义表述方面,通常使用的方法包括：词典定义法、权威定义法、反面定义法、词源定义法、类比定义法、举例定义法、语境定义法、操作性定义法。

学术论文在语篇层面上展开的方法很多,通常可以采用按时间顺序、按步骤顺序、从抽象到具体、从具体到抽象等方法。

要把学术论文写得连贯、流畅,可以使用以下方法：有逻辑地组织论点、使用过渡性词语或表达、重复关键词、使用关键词的同义词等。学术论文还需要注意语言的多样表达,这样可以增强学术论文的可读性。

6. 结果、讨论与结论

论文的结果、讨论与结论部分具备的基本功能是：(1)浓缩观察的事实；(2)归纳试验的过程；(3)分析研究的结果；(4)指出争议的问题；(5)阐明作者的观点；(6)得出最后的结论。

从写作要求的角度来看,结果和讨论部分是论文的关键。在该部分中,作者应当对试验结果的分析和讨论予以充分说明。一方面,试验结果和具体的判断分析通常是逐项讨论的。注意在阐述研究结果时,不要再详细说明试验的过程。另一方面,对试验研究中成功的结果,或结果中成功的部分,应予以评述。同时,对研究中失败的结果,或结果中失败的部分,也应予以说明。最后,结论是作者通过试验、推理得出的最后见解,是整篇论文的归宿,因此要鲜明准确,简短有力。除了对前置的问题给出明确的回应,结论部分还可以对进一步的研究提供意见和建议。

7. 致谢和参考文献

致谢的基本功能是表达作者的谢意,承认他人的工作。

致谢的特点表现在:一是内容有一定的顺序,如致谢对象通常是"先个人、后组织"、"主要者先,次要者后";二是表达比较正式。写致谢时,内容要具体,陈述恰如其分,而且必须征得被致谢者同意。

学术论文中经常会使用一些图表来辅助思想的表达。使用图表时需注意一些问题,如应有和图表相对应的图题、表题及编号。

附录是论文主体的补充项目,但写入正文又可能有损于论文的条理性和逻辑性,因此为了体现整篇论文的完整性,这些材料可以写入附录。

在学术论文写作中,作者常需参考、引用他人的观点、理论、研究成果等,来支持、说明自己的观点、理论。作者应准确标示这些文献的来源,否则就可能会导致剽窃,引起不良后果。而做好参考文献工作,不仅可以交代作者本人研究的科学依据,而且也可表明对他人研究工作的尊重,同时也便于读者查证有关资料。

论文的参考文献一般包括文中引语(in-txt citation)和参考文献列表(list of references)。其具体格式因学科不同而异,不同的刊物对参考文献的写法也有不同的要求。

Unit 2
Mega Structures: Skyscraper

Teaching Guidance

Read the article, pay attention to the **Words and Expressions** and related *sentences* and *paragraphs*.

In the center of New York City, a new vision for the urban landscape emerges. Twenty-two thousand six hundred and eighty tons of steel over eight thousand panels of *crystalline* glass, it is one man's towering dream becoming reality. One blank part is said to be one of the most environmentally friendly skyscrapers in the world. With global temperatures's rising, one green superstructure may alter the way we build our future.

2.1 Energy-saving Systems Combined into Skyscraper

It is 8 am, in the center of Manhattan, and these men are on mission. Today they must *raise over* 45 tons of steel to 220 meters in the air. To complete the 48th floor of this monumental tower, they are in a race to get this skyscraper to its full height.

Worker: All right, Frank. Everything looks good.

Once complete, *Bay Bryant Park* will tower 365 meters above Manhattan streets, second only to the Empire State Building as the tallest skyscraper in New York City. To do that, they must build the structure 288 meters high and then top it off with a 78 meters' spire, one of the largest spires ever erected in the city.

But the building's biggest achievement will not be its height. Bay Bryant Park is said to be one of the greenest skyscrapers in the world. Its combined green features will be unprecedented. It is an ambitious plan to have so many *energy-saving systems* combined into one skyscraper. And it is the *brainchild* of this man. Architect Richard Cook's dream is to raise the most environmentally friendly skyscraper in the world.

Richard Cook: As an architect, I think we must immediately radically change the way we build buildings and radically change the way we build cities. I hope that when people look at Bay Bryant Park, they think that it could be an *icon* of what we started thinking *fundamentally* differently about the environmental impact of a skyscraper.

Cook's vision——Bay Bryant Park is out to tackle a big problem. Buildings *account for* nearly 40% of total US energy consumption. Each of these buildings need energy for lights, elevators, phones, computers, faxes, air conditioners and running water *round-the-clock*. It is a problem of the Bank of America and the Durst Organization, who were ready to confront when they *commissioned* the revolutionary building.

Anne Finucane (Chair of Environmental Commitment, Bank of America): We want to be seen as stewards of the environment. It was clear that we could actually build the most environmentally sound building skyscraper in the world.

The bank gave Richard Cook the opportunity he was looking for. Cook's vision——Bay Bryant Park will combine multiple *cutting-edge* methods to save energy and resources. (1) First is the *ventilation*, a *state-of-the-art* system will take in air, clean it, *circulate* it through the tower, then expel it back out into the city as shown in Fig. 2 – 1. (2) Next is the *recycled* water, an advanced *reclamation system* will collect all rainwater that falls onto its roof, storing it in large gray water *tanks*. The rainfall will be reused throughout the tower to help supply water systems for the skyscraper as shown in Fig. 2 – 2. (3) Then is the glass, a new *crystalline facade* will help keep the heat out, lowering dependence on *interior lighting* and *air conditioning*, saving on electricity. Cook's first challenge is coming up with the initial design.

Figure 2 – 1　The ventilation system　　　　Figure 2 – 2　The recycled water system

Richard Cook: As an architect, you start with the blank *canvas*, and it is what this building should look like. We are really interested in the pursuit of the perfect form whatever it might be at this location at this place. So this board that we are looking at here are a whole series of *sketches* of ideas. And when I look at this, I can see pieces of the building that exist in its final form.

2.2　The Best Way to Get through Rock

Building his vision is not an easy task. Not only is Cook raising the tallest *energy-efficient building* in New York, but also has to build it in one of the toughest places in the world, *Midtown* Manhattan. Erecting a skyscraper in New York City creates extraordinary challenges especially at the corner of 42nd Street and 6th *Avenue*, one of the busiest *intersections* in the city. To start off, the crews must do one of the deepest foundations in Midtown.

Serge Appel (Project Manager): This is an incredible *deep foundation*. It is the full width of the site with 200 feet long, 450 feet wide and it is about a hundred feet deep.

Over one hundred ninety-eight thousand *cubic meters* of earth had to be removed from the site, but the team faced a big problem. *Blasting* in Manhattan is forbidden.

Serge Appel: Typically blasting is the most efficient and quickest way to get through all that rock, but we have *subway terminals* on 42nd Street and another subway line coming up and down 6th Avenue which are right there. And we can't really blast adjacent to that. There is also *historic*

buildings all around the area as well as a 50-storey tower directly adjacent to us, not really conducive to blast it.

So they had to do it by the old-fashioned way, by *digging*.

Serge Appel: You can't *dynamite* Manhattan, so it is all *chiseled* and machined out with large *excavating machines*.

2.3 Steel Beam Made from Recycled Material (Scrap Metal)

All told it took almost a year of excavation. By July 2005, the foundation was finally complete. Almost two years later, the building stood 48 storeys high. That's over 200 meters above the streets. With seven floors left to go, they are almost at the top. And work continues *furiously*. Today, they must lift almost 36 tons of material into the tower with the help of a *crane*, which sits almost 228 meters above the street. At the foot of the skyscraper, the foreman prepares the load. The crane *operator* is so high up that he can't see the loads hundreds of meters below. At this height, the operator must lift each load blind relying solely on the voices of his crew.

The crane's biggest job is lifting the enormous steel beams to *ironworkers*, who continue to build the floors of the tower, but firstly the crew must get them to the site.

The beams arrive on large *flatbed trucks*.

Thomas Kenney (Steel hauler): This (the beam) is just an *average load*. It is about 23 tons.

But getting a large truck through the city traffic can be a nightmare.

Thomas Kenney: We come in like 5:00 in the morning for traffic reasons. Michael calls us and says 'alright come on up', so then we start coming up. But that's the biggest nightmare getting here.

By midday, the traffic gets worse.

Thomas Kenney: I need a long-distance to make a turn. I'm going to have to go down and cut over three lanes to make a left-hand turn down there. There is no room to park on these streets, so that is the issue.

Once the steel arrives, the men work quickly raising about 20 tons of steel to the 48th floor. Although these look like average steel beams, they are actually a secret weapon in this building's "green Arsenal".

To be one of the most environmentally efficient skyscrapers in the world, Bay Bryant Park will embody *state-of-the-art* green systems from surface to core. Architect Richard Cook's conception is about changing how commercial buildings are constructed.

Richard Cook: The fundamental breakthrough wasn't to design a building and then see how green you can make it. But it was in the DNA of the project, it was how we fundamentally *rethink* the way we build a building.

Cook's team *decreed* that many of the materials used to build Bay Bryant Park must be sourced within an 800 kilometer radius of the site, cutting transportation costs and energy use. What's more, the most basic building materials of the tower are largely recycled. It is one way that Cook's team saves energy on the building from the ground up. All of the steel used to build a skyscraper is at least 60 percent *recycled materials*. Every steel beam going into the building started from a plant, which collected *scrap metal*. Then steelworkers melted down, removing impurities and adjusting its chemistry components.

Joe Stratman (Staff from steel plant): Our steel first of all starts with scrap metal. We recycle the scrap metal from almost 95% of our *raw material*, like old cars, old washing machines and old dishwashers. We shred it, compact it, sort it and then that becomes our basic starting point.

This *molten liquid* runs through the moulds and transforms into steel beams. Then machines roll them out onto *cooling beds*, most still glowing from the heat. Once cooled, workers test them for strength and *durability*. From this steel mill, each beam eventually makes its journey to the Bay Bryant Park site in Manhattan (Fig. 2 – 3). But steel is not the only sustainable element that leaves this plant. Even the waste from the steel mill helps create *sustainable materials*. Steel mills produce blast furnace slag, a waste byproduct of steel production that is usually just thrown away. But for the Bay Bryant Park site, this *slag waste* is reused as a key element of the building's concrete, making up 45% of the mix. This leads to a drop in the *carbon emissions* created during concrete production.

(a) (b)

Figure 2 – 3 The recycled steel manafacturing

2.4 Blast Furnace Slag Replaces Regular Cement

Richard Cook: When we work at scale and it's an enormous building, little changes make big differences. Changing from *regular cement* to *blast furnace slag* save 56,000 tons of CO_2 production. So the concept is working in a really big scale by little changes, you will make big difference.

All told, 52,000 cubic meters of this concrete slag mix will fortify the superstructure. Not only does the production of *concrete slag* emit less carbon, it also makes the concrete stronger. Building with green materials is challenging in itself, but the team at Bay Bryant Park must also face all the construction hurdles of any new skyscraper, like the unforeseen forces of nature. At the site, the sky is open, a torrential downpour which would one day fill the building's water reclamation system, now forces most of the construction to a screeching halt.

David Horowitz (Vice President of Tishman Construction Corp & Project Manager): Between the driving rain and the wet decks, it's difficult to get people up here in a safe environment for them to continue the steel *erection*. All the steel erection and the perimeter work is done for the day.

But not everyone is deterred by the rain. To *keep on schedule*, work must continue. The team braves the elements, pouring concrete floors even in the *downpour*. Surprisingly, the rain doesn't affect the concrete strength. The workmen must keep pace with the steel erection to finish the building on time. The building rises, so does the pace of the concrete team.

The next day, the sky is clear and just in time. Today's schedule is lifting one of the buildings largest beams to the rooftop. And now the problem isn't rain. It's the wind. Weighing at *whopping* 9,000 kilograms, this *beam* must go up 268 meters. It will act as a support beam for the spire that will be set on the roof. The crew takes extreme precautions, strapping the beam to the crane. They want to avoid any *catastrophic damage*. Since cranes cannot control their loads laterally, wind poses a real hazard. And today the winds ceaselessly shift. The team is eager to move, but for now they must wait. The men take every precaution because working on a construction site can be deadly dangerous. *Crane accidents* alone have claimed an average of 82 lives in the past 10 years in the United States. In November 2006, an entire crane collapsed in Seattle, killing one man and damaging three neighboring buildings. 8 months earlier, a crane worker plunged to his death in Miami after the safety platform on which he was standing damaged a crane brace. The list goes on. So this mean the men will take no risks.

Michael Keen (Teamster): We're going to send the load back today because as you can see it's too windy. We almost send the beam on 800 feet in the end. But it's too windy. And picking it, we're worried that the load will spin into the crane. So we'll try to get it next morning.

But today they must complete the 52nd floor of the superstructure. It's easier said than done. The crews work over 243 meters up the ground. Each man must climb two 12-meter ladders and walk a maze of steel beams to get into position. To work in such heights, the men must have nerves as strong as the beams on which they walk. It's a *high-wire* balancing act with no safety net. As the cranes *swing* the steel beams over them, the men must carefully position each beam, locking it together like a *jigsaw puzzle*. It's a careful balancing act, but the men are undaunted. They succeed finally. With the final beam, they complete the 52nd floor of the tower. Each day floor by floor, these men bring the superstructure one step closer to the *spire installation*. They must get this building to its full height at fifty four storeys before the spire can be completed. The building continues to rise skyward as many skyscrapers have before it, but its unique green features set it apart. However this tower is not the first green skyscraper ever built, the team learns from techniques used by other architects and engineers.

Richard Cook: We evolved to make buildings better and better. This is an evolutionary process. We could not do this if others hadn't been before us.

2.5 Utilizing Natural Air & Green Space

A key green feature in this building is its use of natural air and *green spaces*. They want to use it in a new way based on *concepts* used by other green skyscrapers. One of the first skyscrapers to use natural air flow in its design is the Commerz Bank Tower in Frankfurt Germany. Built in 1997, it's the third tallest building in Europe. Its use of natural winds and green spaces influence the green skyscrapers followed, including Bay Bryant Park. The *Commerce Bank* uses natural air flow to make this tower more energy efficient. A *central atrium* runs up from the *lobby* with 190 meters to the top of the building without a roof to top it off, this shaft acts as a natural *ventilator*. Winds blow upward through the offices to the people inside, as shown in Fig. 2 – 4. This design created a complex construction challenge. The atrium forced the *core support* of the building, which is traditionally in the center of the structure to the outer perimeter. The design keeps the building cooler. The Commerz Bank Tower also uses nine sky gardens and patches of green life that help cool the structure, taking in carbon dioxide and emitting more oxygen.

Figure 2 – 4 The wind blows upward through the offices to the people inside

Stephan Behling: On the one hand, it's the social benefit, the gardens are places to meet. On the other hand they're the *lungs* of the building.

The building is 80% naturally ventilated, saving 30% on energy use.

Stephan Behling: If you do not have to switch on a machine, it's the most efficient machine. So if you do not need air conditioning, and you can design a building in a way that you don't need this incredibly energy *intensive* machine running, that is a major breakthrough.

Richard Cook and his team took these concepts of natural air flow and green spaces a step further. At Bay Bryant Park, air enters the building and the top floors, then passes through a giant air *filtration* system, as shown in Fig. 2 – 5. The skyscraper will not only clean air for itself, it will also clean air of outside areas immediately around it.

Richard Cook: And we *filter out* 95% of the particulate. We supply it to everyone and when it exits the building, it will be cleaner than when it comes in. So you could think of this block of Man-

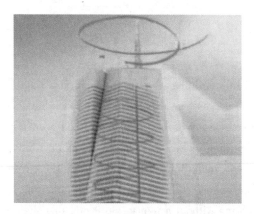

Figure 2-5 The air filtration system in Bay Bryant Park

hattan operating as a giant *air filter* for this part of the city as the air exits back out into the city.

2.6 Roof Cooler, Gray Water of Capturing & Reusing

Cook also wanted to use green roofs at Bay Bryant Park, just as he's done at his own offices.

Richard Cook: In our cities, what happens is something called a *heat island effect*. The roof of my office was 170 degrees last August, which was hot bubbling tar beach. And by planting these bags, we radically reduce the temperature of our roof, lowering our own little piece of New York City, lowering its temperature.

The green roof at Bay Bryant Park will not only keep the roof cooler, it will also serve as one of the collection points for rainwater. Cook and his team want to reuse every drop of rain that hits the building's roof (Fig. 2-6).

Figure 2-6 The green roof at Bay Bryant Park

Serge Appel: We're standing on top of the 6th floor roof, which is one of the main roof services where we're collecting water. Basically, this entire area is going to be covered in a green roof.

When rain hits this roof, it's absorbed by the plants, the *excess funnels* down into the building. There will be four tanks placed on different floors throughout the skyscraper and one in the *cellar*. This *gray water* system captures and reuses almost all rainwater and most wastewater, saving over

400 million liters per year, which translates into an annual saving of about a half million dollars.

For Cook's system to be operational, the crew must first get this tank into place. For example, the men must lift one of five tanks, a 32,000-liter tank, into place from the street level to the 41st floor above. It's bulky and weighs 9000 kilograms. So getting it into the building is a formidable task. There is another risky operation for the crew. The *straps* must hold until the tank reaches its destination——over 122 meters above the streets. Up on the 41st floor, the men guide the giant tank from the crane onto the building. But its large size makes that difficult. The men must resort to using basic tools like chains and rope to do the job. Once in place, the gray water storage tanks will hold almost 260 thousand liters of water combined. Deep in the building cellar, Cook's team has installed a unique system——the *chilled water* by night to efficiently cool the building by day. 24 meters below the street surface sits the largest collection of cooling tanks in the city, with 44 in all. These 7,570-liter tanks freeze 227 thousand kilograms of ice each night. By day, this ice helps keep the building cool by feeding the air-conditioning system for the entire skyscraper, ensuring that it all stays energy efficient and cost-effective.

2.7 Allowing Light to Enter While Blocking out Heat

The goal of the Bay Bryant Park Tower is to reduce energy consumption from the grid, by half that of a building of comparable size. But it's an uphill battle, because the city keeps getting hotter. Traditional skyscrapers not only use a lot of energy, but also radiate heat into the air.

Stuart Gaffin (PhD., Center for Climate Systems, Columbia University): Urban scientists sometimes use a term called the *urban canyons*, which is a kind of a version of a natural canyon, although it's man-made. Mostly they add to the heat effect, couple of ways of which is they increase the surface area in a city that's exposed to sunlight. So we have many more times of the area of Manhattan that is exposed to sunlight than just the surface area of the island.

The Empire State building's facade is primarily made of 5663 cubic meters of *limestone* and *granite*. The sheer size of the building also helps to heat up the city.

Stuart Gaffin: The Empire State Building once I estimated to have about 40 times the surface area of the block it sits on. And that extra surface area is available to absorb sunlight in the morning and in the afternoon during the day. And this is a very real effect in terms of cities absorbing more sunlight and causing them to have higher temperatures.

On this 29.5 degrees *Celsius* day, infrared photography shows surface areas on the Empire State Building highlighted in dark red, which radiates heat over 38 degrees Celsius.

Stuart Gaffin: Just about everything we do tends to add heat to our cities. We can't remove buildings. Now this is a really tough problem. Are we able to avert these high levels?

In the past, most buildings have not been designed to address *global warming*, but it's about to change. Now a few architects are designing buildings with glass *facades* that operate more efficiently. Cook wants Bay Bryant Park's facade to keep interiors cooler. And one of its *predecessors* does much the same with its modern tower. Key to its efficiency is its use of glass and steel. One of the most distinctive facades in New York stands on Eighth Avenue between 56th and 57th Street. At the Hearst tower, a traditional 86-year-old facade meets the cutting edge in ultra-modern design. Not only is it modern, it's also one of the greenest skyscrapers in existence. The Hearst tower uses state-of-the-art window technology in its glass façade. Its special glass coating lets *visible light* through while keeping heat out, interior state cooler and demands less energy for air-conditioning. In addition, the glass's face allows more natural light to enter office spaces, reducing the need for artificial light and electricity. This facade structure also requires 20% less steel. The new tower's face uses a unique *diagonal design*. Each *diagonal section* spans four storeys and each of the triangles carries the *gravity load* while also providing *lateral strength*, so beams don't need *additional bracing*. In the end, this saves over 1,800 tons of steel, compared to a traditional structure of the same size. But building it was a challenge.

Frank Gramarossa (Project Executive, Turner Construction): This particular job is that there were no *vertical columns*. All columns were with diagonals, and they were actually not two storeys long, there were four storeys long. So we had to figure out how we were going to erect these columns, how to stand up a column and have it physically stand there without bracing it or supporting it temporarily until you could tie the other column in.

It's all worth it in the end. With its cooling systems and decreased use of artificial light, the building consumes 26 percent less energy than a skyscraper of comparable size. Now, Bay Bryant Park is set to outdo that. Richard Cook wanted his skyscraper to save 50% on energy use. As in the Hearst tower, Cook wanted to save energy at Bay Bryant Park by using a state-of-the-art glass *window façade*. Its special coating allows light to enter while blocking out heat and engineers took it a step further.

Serge Appel: We need to find another way to reduce the amount of sun energy that's entering into the building. So at the top and at the bottom of each piece of glass, we have this thing called a Frick pattern, which is baked right into the glasses essentially. And it's a series of *dots* that fade from bottom up and from the top down. If we look right here, we can see these dots on the glass, that's the *ceramic frame* as shown in Fig. 2-7.

The dots are so small that they're hardly visible, but they reflect out the harmful effects of the

Figure 2-7 The black dots on the glass

sun, minimizing heat entry into the building. This keeps building interiors cooler, therefore less energy is needed for air-conditioning. But in order for this facade to be successful, this *glass curtain wall* must first withstand tough conditions. Windows on other modern skyscrapers have made headlines before, sometimes for the wrong reasons. In 1999, windows from a skyscraper in Chicago came crashing down, taking the life of a young mother. During the construction of the Hancock Tower in Boston in the early 1970s, a *violent storm* blew out at least 63 *glass panels*, sending them shattering to the streets below. To avoid any *catastrophes* at Bay Bryant Park, the facade design is rigorously tested before it's installed. At this facility in Miami, Richard Sembello (Director, Construction Research University) and his team tested equivalent glass facades under extreme conditions. They tested the aluminum of glass curtain wall system with high force of winds and rains. A water rack provided the *equivalent* of 20 centimeters per hour rainfall, and a *turboprop* jet engine breathed in 80 kilometers per hour wind force.

Richard Sembello: We're inspecting the wall for *leakage*. We will look along the *gaskets* down, the gaskets at the corners. The wind gusts are also moving the glass in and out, somewhere around the quarter of an inch.

The approved wall system design is then manufactured and installed at Bay Bryant Park. It will take over 8,600 glass panels to sheet the building. Each section stands between four point five and six meters high, and weighs about 453 kilograms. The men must lock each and every panel in place one by one. To do that at these heights is time-consuming and dangerous. Here on the 38th floor, the men hang out over the ledge of the building about 183 meters off the ground. The crews must race to keep up with the ever rising steel structure.

Daniel Tischler (Window Installer): There are two people on the bottom that hoist the panels out. The three men on the top set the panel on the top and set it to the height and then correct in and out in the rack. It takes approximately one week to do one floor, depending on the weather, the floor conditions.

To stay on schedule, they must complete the windows of this floor on time. Just as in any skyscraper construction, time is money and there is a constant pressure on the team to raise the structure on the schedule. But in a green building like this one, there is even more at stake. Building *sustainable skyscrapers* costs about five percent more than a *conventional structure*. And Bay Bryant Park translates into roughly sixty million dollars. For many companies, these additional costs may be the biggest *stumbling block* against going green.

Stuart Gaffin: Right now, a lot of these technologies are expensive. They're *novel* and they're not really cost effective yet.

But Cook thinks looking only at the *initial costs* is *short-sighted*.

Richard Cook: If we save energy ultimately in the long run, they will save vast quantities of money. Building will radically reduce about half of the energy that it would have used in a conventional billing, about four million dollars a year at today's dollars.

Paul Goldberger (Architecture Crtic, New York Magazine): People are going to be forced to think about *sustainable design* more seriously whether or not they want to, because they're going to be forced by their pocket books. The price of electricity is forcing people to be much more *environmentally sensitive*.

This is where *business interests* and *environmental conscience* converge. And when a commercial bank decides to go green, perhaps others will follow.

Anne Finucane: In the short-term, it is true that there were some *additional costs* associated with the building. However, from an energy use perspective, from the health of our *employees's* perspective, from the *perspective* of long-term play, this is a *homerun*.

Back on the 38th floor, the window installation is almost complete. The men hang precariously over the edge of the structure, pulling together the final panels of this floor. As the windows continue to rise up the skyscraper at breakneck speed, the steel structure will soon stand at its full height. It's July 2007, almost three years after the skyscraper broke ground, and the men prepare for a monumental occasion. They're about to lift the final beam that will complete the top floor of the superstructure.

Serge Appel: It is a great honor to participate in this project. We are proud to stand here. The more beautiful these buildings are, the more honored we are to participate in them.

They leave their signatures as a *permanent mark* on the superstructure.

Daniel Tischler: Proud, we built this building.

2.8 Infinite Imagination of Green Building in the Future

The final beam reaches the top floor, but the work is not over yet. Now with the floors of the superstructure at its full height of 288 meters, the men move into their next big challenge, setting a 78 meter *spire* on top of the giant steel structure.

After weeks of waiting, the time has come. But just as the men are beginning the spire installation, something goes terribly wrong.

News: A bucket fell about 53 storeys off a crane in Midtown, breaking several windows of the Bank of America Tower under construction.

A container has crashed into the building, smashing windows and sending glass shards and materials, hurtling down almost 245 meters to the streets below.

Passerby A: I heard a loud voice, so all the stuff falling down.

Passerby B: The bucket fell down and hit part of the building to put part of the building with it, and some of the *debris* might dropped out and actually hit the windows of the building.

Since the shattered windows are adjacent to the *crane struts*, the emergency crews are set up to determine if the crane is still structurally sound, and to clear broken glass out of the damaged window *frames*. Four *pedestrians* and four workers are injured, the accidents shuts down a two-block radius around the site. All construction comes to a complete stop and the team must regroup. It takes nearly two weeks to address all the safety concerns. Then finally, the men move on to the last big job at hand——topping off the Bay Bryant Park skyscraper. The crews must raise a giant spire at top of the steel frame. Only then will the main structure of the building truly stand at its *apex* of 365 meters. The spire is so large that it must be transported in pieces on several trailer trucks and massive loads going into Manhattan need special care.

Michael Keen (Steel Hauler): Going into New York City basically lost two days because you have to go at night. Then first thing in the morning is to get the site to unload the cargo. It is true what they say is that the city that never sleeps. It gets a little dangerous at times but you have an escort behind you, which can block it off a little bit.

Early the next morning, the spire pieces finally arrived at the site.

The spire is made up of 70 individual steel pieces. Each must be lifted to the rooftop where they are assembled and attached to one another. The process is expected to take nearly five weeks under ideal weather conditions. Despite the fog, the men get to work. The crane lifts a 7.5-meter long of at least one and a half tons section into position. At almost 275 meters in the air, a slight miscalculation can prove fatal. Iron workers carefully guide it into place, then weld and bolt the sections together. The operation goes off without a hitch. Over the next few weeks, the spire continues to go up section by section. In the end, Bay Bryant Park tower will stand with 365 meters high. When it is fully operational, the tower will be one of the greenest in the world. And it may be leading the way into our urban future. The building is expected to reduce both energy and water consumption by 50%, and its occupants will benefit as well.

Anne Finucane: They will have a *healthier* environment. They will have clean air to breathe. They will be in an environment that they can feel not only proud but we will be giving them a better physical environment to be in and this is a good *blueprint* for the future.

But with the high cost of building green and with global temperature's rising, can Bay Bryant Park make an impact?

Stuart Gaffin: By itself, usually one building won't make a difference. We've really got to get this widespread because Manhattan is a small island, but we really have to transform a huge numbers of surfaces, and systems and technologies in order to make a dent in the climate.

Some believe the *momentum* has already begun to shift. And the trend may be contagious.

Paul Goldberger: Many of the ideas in Bay Bryant Park are finding their way into other buildings already. And in another few years, we'll see lots of things like that.

Richard Cook: I don't think there isn't any question but that is what we're at a *tipping point*. Maybe we've crossed it and we're already over the line. The goal was to create the greenest building we possibly could. So people are watching carefully and thinking how did the Bank of America decide to spend the money to do a seriously green building? And in that small way, I hope that the building will be a *trigger* for lots of other projects around the country and around the world.

But Cook's vision at Bay Bryant Park is among the first steps towards significant change. Green architecture and skyscrapers are constantly evolving. It's not an easy process, but it's a vitally important one for the growth of our cities and the future of our planet. So as far as green buildings go, it seems at least for now, sky is the limit.

Words and Expressions

crystalline   adj. 水晶的，似水晶的
brainchild   n. 独创的观念，点子
icon   n. 偶像，崇拜对象
fundamentally   adv. 从根本上，根本地
commission   v. 委任，授予
cutting-edge   adj. 前沿的，最前沿的
ventilation   n. 空气流通，通风方法
circulate   v. （使）循环，（使）流通
recycle   v. 回收利用，使再循环
tank   n. 油水箱，贮水池
canvas   n. 油画（布）
sketch   n. 草图
midtown   adj. 市中心区的
avenue   n. （南北向）街道
intersection   n. 十字路口，交叉路口
blasting   n. 爆破
dig   v. 控，掘
dynamite   v. 破坏，炸毁
chisel   v. 凿
furiously   adv. 猛烈地，气势汹汹地
crane   n. 吊车，起重机
operator   n. 操作员
ironworker   n. 铁工厂工人
state-of-the-art   adj. 使用最先进技术的
rethink   v. 重新考虑或再想
decree   v. 命令
durability   n. 耐久性，持久性
erection   n. 安装
downpour   n. 倾盆大雨
whop   v. 打
high-wire   adj. 高空走钢丝的，危险的
swing   v. （使）摇摆，（使）摇荡
concept   n. 观念，概念
lobby   n. 门厅，大厅
ventilator   n. 通风设备
lung   n. 肺

intensive   adj. 强烈的
filtration   n. 过滤，滤清
excess   n. 多余，过量
funnel   v. 便成漏斗形
cellar   n. 地下室
strap   n. 带子
limestone   n. 石灰石
granite   n. 花岗岩
Celsius   n. 摄氏；adj. 摄氏的
facade   n. 建筑物的正面
predecessor   n. 原有事物，前身
dot   n. 点，小圆点
catastrophe   n. 大灾难
equivalent   adj. 相等的，相当的
turboprop   n. 涡轮螺旋桨发动机
leakage   n. 漏，漏出
gasket   n. 垫圈，衬垫
novel   adj. 新奇的
short-sighted   adj. 目光短浅的
employee   n. 雇工，雇员
perspective   n. 观点，角度
homerun   n. 全垒打
spire   n. 尖塔
debris   n. 碎片
frame   n. 边框
pedestrian   n. 行人
apex   n. 顶尖顶角
healthier   adj. 更健康的（healthy 的比较级）
blueprint   n. 蓝图
momentum   n. 势头
trigger   n. 引发其他事件的一件事，触发
Bay Bryant Park   湾布莱恩公园大厦
raise over   提升
energy-saving system   节能系统
account for   （在数量、比例上）占

round the clock　一整天，昼夜不停
state-of-the-art　最先进的，最新水平的
reclamation system　回收系统
crystalline façade　水晶立面
interior lighting　室内照明
air conditioning　空调
energy-efficient building　节能建筑
deep foundation　深基础
cubic meter　立方米
subway terminal　地铁终点站
historic building　历史建筑
excavating machine　挖掘机
flatbed truck　平板货车
average load　平均负荷
recycled material　再生材料
scrap metal　废金属
raw material　原材料
molten liquid　熔融液
cooling bed　冷却台
sustainable materials　永续材料
slag waste　炉渣废料
carbon emission　碳排放
regular cement　普通水泥
blast furnace slag　高炉矿渣
concrete slag　混凝土矿渣
keep on schedule　按计划进行
catastrophic damage　灾难性损害
wind blowing　刮风
crane accident　起重机事故
jigsaw puzzle　拼图游戏
spire installation　塔尖安装
green space　绿色空间
commerce bank　商业银行

central atrium　中央中庭
core support　核心支撑
filter out　过滤
air filter　空气过滤器
heat island effect　热岛效应
gray water　灰水
chilled water　冷水
urban canyon　都会峡谷
global warming　全球变暖
visible light　可见光
diagonal design　斜交格构设计
diagonal section　斜交组件
gravity load　重力荷载
lateral strength　横向强度
additional bracing　附加支撑
vertical column　立柱
window facade　玻璃窗立面
ceramic frame　陶瓷熔块釉
glass curtain wall　玻璃幕墙
violent storm　狂风暴雨
glass panel　玻璃嵌板
sustainable skyscraper　永续的摩天楼
conventional structure　常规结构
stumbling block　绊脚石
initial cost　初期成本
sustainable design　可持续设计
environmentally sensitive　环境警觉的
business interest　商业利益
environmental conscience　环保良心
additional cost　额外成本
permanent mark　永久标记
crane strut　起重机支架
tipping point　引爆点

Activities——Discussion, Speaking & Writing

1. Listening practice

　　Watch the video clips, write down the key words you have heard, retell the main idea in your own words and then check your writing against the Chinese scripts below.

Exercise 1：通风

Reference Script：首先是通风，采用最顶尖系统吸入空气，净化之后再对整栋大楼进行循环，再重新排放至市区。

Exercise 2：循环用水

Reference Script：其次是循环用水，先进的回收系统会收集落在屋顶的雨水，贮存在庞大的废水箱。整栋大楼将再利用雨水协助供应摩天楼的用水系统。

Exercise 3：钢梁来源于可持续建材

Reference Script：首先，钢梁来源于废五金，对其进行再利用，将近95%的原料不过是回收利用的废五金，如旧车、旧洗衣机、旧洗碗机。我们绞碎、压缩、分类，基本上就可以开始制造钢梁了。融化的液体流经模子，转化为一根根钢梁。接着机器把钢梁推上冷却台，这时大多数钢梁仍因为高热而发光。冷却之后，由工人测试其强度和耐久性。

Exercise 4：高炉矿渣的回收利用

Reference Script：炼钢厂会产生高炉矿渣，这种炼钢产生的附属废料通常会直接丢弃，但湾布莱恩公园大厦的工地把废弃的矿渣回收利用，作为大厦混凝土的重要组成部分，占拌合料的45%，因而降低了混凝土生产过程中的碳排放。

Exercise 5：利用自然气流

Reference Script：中央中庭从大厅向上延伸190m到达大楼顶部，没有屋顶的覆盖，这个天井充作自然通风装置，风从上通过办公室，吹向里面的人。

Exercise 6：大厦净化空气装置

Reference Script：在湾布莱恩公园大厦，空气进入大厦和顶部楼层，再通过一个巨大的空气过滤系统。这栋摩天楼不只能够净化大厦内部的空气，也能净化周围一带的空气。

Exercise 7：废水收集再利用

Reference Script：雨水落在屋顶上时就会被植物吸收，多余的雨水汇集到下面的大楼。整栋摩天楼在不同楼层分置四个水箱，在地下室也有一个。几乎所有雨水和大多的废水都由这废水系统收集再利用，每年可以省下四亿升的水，相当于每年节省50万美元。

2. Group-work: Presentation

Group (5 to 7 members) discuss and give your opinions about the **Green Building**. Choose one theme below for your presentation. Clearly deliver your points of the following themes to audiences. Each group collects information, searches for literature, pictures, etc. to support your points. Prepare some PowerPoint slides.

10 minutes per group (each member should cover your part at least one or two minutes).

NEED practice (individually and together)!! **Gesture** and **eye contact**. **Smile** is always Key!!

Themes on Green Buildings

(1) New Development of Green Architecture in Big Data Era
大数据时代绿色建筑新发展

(2) Intelligence and Digital Technology of Green Buildings
绿色建筑智能化与数字技术

(3) Green building and Indoor Environment Optimization
绿色建筑与室内环境优化

(4) Energy-saving Renovation Technology and Engineering Practice of Existing Buildings
既有建筑节能改造技术及工程实践

(5) Energy-saving Operation and Energy Consumption Control of Large Public Buildings
大型公共建筑的节能运行与能耗控制

(6) Research on Comprehensive Energy-saving Technology of Green Building Materials and Envelope Structure
绿色建材与外围护结构综合节能技术研究

(7) Application of New Energy and Green Building
新能源与绿色建筑应用

(8) Resource Utilization of Construction Waste
建筑废弃物的资源化利用

(9) Green Construction Technology and Demonstration
绿色施工技术与示范

(10) Evaluation and Construction of Green Low-carbon Eco-city Zone
绿色低碳生态城区评价与建设

(11) Green Building and Sponge City Practice
绿色建筑和海绵城市实践

3. Independent work: Writing

Search for **Green Building** in relevant Index, like Engineering Index(EI), Scientific Citation Index(SCI) and China National Knowledge Infrastructure(CNKI). Then write a report independently on **the Green Buildings in China.**

Unit 3
Chinese Super Bridges

Teaching Guidance

Read articles, pay attention to the **Words and Expressions** and related **sentences** and **paragraphs**.

This article will introduce three world-class bridges, including the arch bridge of Lupu Bridge, the suspension bridge of Runyang Bridge and the longest Sea-Crossing bridge in the world of Hong Kong-Zhuhai-Macao Bridge, all made in China.

The super *arch bridge*, the *spectacular suspension bridge*, and the longest sea-crossing bridge in the world were designed to conquer the *breathtaking* distance. By withstanding typhoons and earthquakes, they passed the greatest bridge-building boom ever, in the planet of the fastest growing nation, China. What is taken to build world's impossible bridges?

3.1 The World's Longest Arch Bridge in 2003

China has two world-class longest rivers, dividing the country and disrupting traffic by land, which are a bridge builder's *nightmare*. The waterway is busy with shipping traffic, which can not be delayed. The rivers are swift, deep and wide, the mud may *submerge* the bridge foundations. Getting worse, like typhoons of 240 kilometers per hour or earthquake almost striking at any time with even seven point zero at Richter scale, these challenges battled Chinese engineers for centuries. But they have been meeting them head on, creating one of the most spectacular *bridge spans*. One of the most daring bridge is Shanghai's Lupu Bridge, the longest arch bridge in the world by 2003. 34 million kilograms of steel, standing on the busy Huangpu River. The Lupu's arch rises more than 90 meters, 120 *cables* hanging from the arch to hold the roadway. The arch spans the entire river of 550 meters wide with no support below. It's standing today as a *masterpiece* of bridge *construction*. But to create the bridge of this kind and this size was a *daunting* challenge, it was not designed until the city was desperate for a bridge.

In 1999, Shanghai's population approached 13 million, as China's largest city was bursting, fleet of cars *clogged* streets. City planners needed to ease *congestion*, but the Huangpu River was in the way. It completely divided the city in half, a new bridge would help Shanghai grow, but building one would be a *major* challenge. The river was nearly half a kilometer wide. A new bridge would need to be designed, and it was not to sink into the Huangpu's boggy soil or block one of the busiest rivers in China. The ship traffic would have to keep flowing.

There was one cost-effective solution, a *cable-stayed bridge*. In this design, the roadway *suspended* by cables, which attached directly to enormous *cable towers*. The cable-stayed bridge could easily stretch across the Huangpu River, but Shanghai already has three cable-stayed bridges. City officials wanted something special, more than just a bridge, they wanted it a monument. They made a *radical* decision, Shanghai would build an arch bridge, one biggest of any *unprecedented*.

We'll be easier to build a cable-stayed bridge, but Chinese always like arch bridges, Chinese has been building arch bridges for 1500 years. In the Lupu's design, the roadway would hang by cables attached to the arch. Without the arch, the roadway would sink and *collapse* under own weight, but an arch doesn't sink, the downward forces tried to make the arch spread out. So as long as the basis of the arch could be kept in place, the arch would keep steady. For a quarter of century, the New River Gorge Bridge in West Virginia held the record for the largest arch bridge in the world. This record has never been challenged before, because building this big bridge was difficult and dangerous, and building even bigger will *magnify* every problem, but Chinese would not hold back.

3.2 The Challenge of Building the Arch

In October 2000, construction of Lupu Bridge began, workers drove in the *piles* that supported the ends of the bridge. Steel workers forged the first of the arch's 64 *segments*. But the design was extremely complex, two equal arches were *crooking* inwards. Workers had to make each segment to different *specifications*, depending on the place in the bridge. When each segment arrived on the site, giant *hoist* slowly arose them into position. This work was tricky, some segments weighed as much as 150,000 kilograms. Every time workers installed a new segment, the hoist was to slide forward directly above, ready to lift the next segment. As the arch grew, engineers faced challenge of keeping it upright. Without the support, the two sides would fall into the river. But *scaffolding* would block the ship channel. The builder's solution was to hold the arch from above(Fig. 3 – 1). Firstly, workers erected temporary towers of more than 120 meters high at each end of bridge. They attached steel cables from the tower to the arch, two cables supported each arch segment. The arch arose piece by piece toward the middle, additional cables kept towers from *toppling*. The bracing was so extensive that it was almost like building the second bridge. The cable-stayed bridge would be *dismantled* immediately after the arch bridge was completed. But until the arch was finished, this was a *precarious* system, with each step the arch growing increasingly *vulnerable*.

"An arch is only an available *structural* system after the ends of the arch are connected together. The two ends are not only unstable in vertical direction because there is no compression,

Figure 3 – 1 The solution to holding the arch

but also they're very *susceptible* to the wind." The structural engineer Mark Ketchum said. The builders had to try to close the arch as soon as possible, but it was slowly going since the design called for the bridge segments to be *welded* together from inside. The space was *cramped* and the *ventilation* was poor, the *welding torch* heated the air to 60 degrees Celsius. It was so hot that the welding workers worked only 15 minutes at a time. To complete the bridge, they needed to make over 300 kilometers'welds. Outside surveyors made sure the arch stayed on track. Gravity, temperature and wind always worked, changing the size and position. If the engineers miscalculated, the final segments would not fit, meaning the arch could not be closed. Instead, it would be left dangerously exposed, and typhoon could strike at any moment. At least two typhoons hit this region every year. The winds of a major typhoon could easily *wrench* open weak *joints*, shake apart a support tower or *snap* cables. The weight of the arch would do the rest, the entire bridge could *plunge* into the river.

Before building began, the Lupu Bridge's designers tested their model bridge against the force of a *magnitude* 7 earthquake, and 270 kilometers per hour typhoon winds. The tests predicted that the arch could survive nature's worst *ordeal*, even when it was not completed. But builders would not sleep soundly until closing the arch. The budget, schedule and safety of the entire project, all lied on if the bridge arch can be closed, and the bigger the arch, the more chances to go wrong.

Mark Ketchum: Anytime you build a bridge out of segments piece by piece, there is a chance that the last piece to put together in the middle is not perfectly fit.

After going well for months, construction appeared to slow down. "They had a bit trouble at the end to install the last piece segment. So it stopped for a while." The Lupu's senior designer explained the reason. But the bridge was still set, unfinished and fragile, and a *destructive* typhoon was approaching.

By July 2002, almost two years of construction, China's Lupu Bridge stood over Shanghai skyline, but it was still not completed. The bridge was now at the most vulnerable and the worst possible moment. Typhoon Rammasun was growing in the Pacific, just 130 kilometers away from the city. For 24 hours, the storm *pounded* the Chinese coast, the high winds knocked down the trees, damaged homes, and blew the unfinished Lupu Bridge. Five people in the city were killed, then the typhoon turned to the north and disappeared. This time, the Lupu's support held, the bridge was *unscathed*. But the builders knew they might not be so lucky next time, so they needed to close the arch before the next typhoon hit. The engineers finally had the arch's last section ready for installation, the final piece began the journey to the top. The builders could not be certain if it would fit perfectly, even after they slid it to place. Workers quickly welded one side of section, but on the other side, the gap was too wide. They needed to wait till the outside temperature rose, so that the steel bridge arch expanded to close the gap. At exactly 20 degrees Celsius, the gap did close, this last joint was *sealed* with permanent *bolts*. Finally it was time to celebrate, but the giant arch was not yet out of danger. Engineers still needed to prevent it from expansion outward, getting into the soft soil, and collapsing.

3.3 The Tied Arch to Prevent Collapse

Arch bridges like the New River Gorge are *supported* each end by solid *rocks*. The loads of the heavy arch pushes outward against the rock, but the stone does not budge and that force keeps the bridge erect. But the soil of the Huangpu river was too soft to resist the Lupu's weight, the ends would slide apart, as in Fig. 3 – 2, the arch would collapse, unless the engineers lined the ends together.

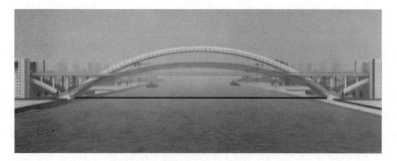

Fig. 3 – 2 The ends of the arch might slide apart

This is known as a *tied arch*, the builders stretched 16 cables between the arch basis, eight cables would remain visible after the bridge was finished, the other eight would be hidden inside the road's spans. Together these cables would act like the *strings* of the arch's bow (Fig. 3 – 3), without their constant force, the arch would be *flattened*.

Man-Chung Tang (邓文中): When there is load on the arch, it will create a very high *horizontal* force, pushing the ends of the bridge away, so you put a cable to tie the ends together, so these horizontal forces would *eliminate* each other in the cable.

Maintenance crews would need to keep eyes on the cables as long as the bridge's standing, they would regularly test *tension* to make sure it stayed about 2,200 tons. The Lupu arch was ready to carry load of the roadway. Workers began to raise 15 *road deck* sections, which was another heavy job that some sections weighed 450,000 kilograms. At this time, the Huangpu River

(a)   (b)

Fig. 3 – 3 The cables between the arch basis acting like the string of the arch bow

was not an *obstacle*, it was an asset, which allowed the enormous bridge span be *floated* to the construction site on boats, it was far easier to float than on land.

On 28 June 2003, the Lupu Bridge opened, two and a half years after construction started. The *contractors* met the deadline in spite of delays. Shanghai now had a *showpiece*, as well as an important new transportation route, and China had its first record for setting the span, the longest arch bridge ever built. But this victory was quickly shadowed, by China's *ambitious* needs for bridges. Chinese wanted more bridges to link the entire nation. It was hard enough to conquer the Huangpu River, which *sliced* Shanghai half, but China had another river that cut the whole country in half.

3.4 The World-Class Suspension Bridge

The Yangtze is China's longest river and the third longest in the world, it was a *watershed* in China's geography and economy, a 6300-kilometer obstacle to China's growth. Three hours west of Shanghai, the Chinese began to bridge the divide with one of the longest mega spans, a suspension bridge called Runyang. The Runyang Bridge needed to be three times further than Shanghai's Lupu Bridge, more than a kilometer in a single bound. At that time, it would be the longest bridge that Chinese ever built, and the third longest suspension bridge in the world. Its enormous twin towers *would climb* over 200 meters into the air. For residents on Yangtze's north side, the bridge would mean faster and safer travel to Shanghai, and was a chance to share China's new wealth. The north bank was cut from Shanghai by the Yangtze, but now the Runyang Bridge would replace a *ferry* ride that took two hours by a 10-minute drive.

3.5 Big Challenges: Ships, Typhoons & Earthquakes

In 1998, designers began to consider an idea about building a bridge in Yangtze between the ancient cities of Yangzhou and Zhenjiang, but immediately they *encountered* a problem. Building a bridge directly between the two cities was impossible, because the city itself extended to the river banks, there was no extra room to add a bridge. Therefore, the designers were forced to build the bridge to the west, but here the gap between the shores was more than doubled. Building the bridge here would demand three mega structures——A 400-meter bridge to the north, an *elevated highway* across the small island, and a massive bridge to the south. In Yangtze, it was difficult to build one bridge, let alone two. The water flow was swift and the soil was unstable, and every summer the Yangtze often flooded with *disastrous* results. A *deluge* happened just months before construction started, *inundating* areas larger than England. If there was a severe flood during the construction of Runyang Bridge, it could *crumble* the project.

The builders tackled the northern bridge first, with a short distance to make job easier. They created a cable-stayed bridge, the cheapest and simplest design. The below roadway was supported by cables attached directly to two *pylons*. This 400-meter structure had no trouble spanning the northern channel. Cable-stayed bridges had ever been built that was twice as long as this here, thus it was a huge construction project, but not broken any records. The state-of-the-art design was amazing both day and night. But building the southern bridge faced far greater challenges. The distance they had to build across was almost four times as far, too far for a cable-stayed bridge to reach without putting several towers in the river. And engineers could not build any bridge towers in the Yangtze River. The southern channel was the main shipping road around the island, there was constant flowing boats. If just one vessel *deviated* the course and *ran upon* a pier, the damage would potentially be catastrophic. Engineers learned this in a hard way, when the ship collision destroyed the bridge and lives in the past. Chinese engineers faced a long list of challenges to connect the two sides of the Yangtze river, the water was over a kilometer wide, and typhoons and earthquakes could strike at any time, but by far the biggest threat to bridge was the unpredictable ship.

3.6　Key Approach of Fixing the Main Cables

In Florida in 1980, the 20,000 ton Summit Venture *destroyed* Sunshine Skyway Bridge. The ship was *navigating* in Florida Tampa Bay in bad weather, it lost radar, deviated off course, and *slammed* into a bridge support, leading the central span collapsed. The crash killed 35 people, including 26 travelers on the bus that fell off the bridge. The Runyang Bridge's builders did not intend to risk such a tragedy, they would base as much bridge on land as they could. And only one design could leap the huge distance, and kept the tower near the shore——a suspension bridge. The design of Runyang, like every modern suspension bridge, relied on two main cables that stretched from one shore to the other. At each end, the one-meter thick cables were rigidly *anchored*, but everywhere else, the *main cables* could actually remove, even across the top of the towers. The towers were like tent *poles*, holding up the cables, but the cables were merely hung from the top of the tower. If the main cables were anchored to the tower top, the towers might not be able to withstand the *pulling forces*, that they would tumble inwards, and the entire structure would be lost. Hanging down from the main cables with 360 suspended cables, these would hold the bridge spans. However, the joints on the main cables were also removable that this flexibility would help the bridge to survive strong striking of the earthquakes and the typhoons. If the Runyang could be built as the design, it would be the third longest suspension bridge in the world at that time——the towers would correspond to 70 *storeys*, and the span would stretch almost 1.5 kilometers. Only one tower was sitting on the Yangtze river. By 2001, just two other bridges in the world had longer spans, which are Denmark's Great Belt and Japan's Akashi Kaikyo.

The construction of Runyang Bridge began in February 2001. The builders had five years to complete the project. The first step was the bridge towers, because the towers needed to withstand the *tremendous* downward forces from the cables, engineers wanted to put them on the *bedrock*. But the Yangtze mud was *incredibly* thick, to reach a solid *footing*, builders needed to sink 32 piles at each tower site. These piles were like giant *stilts*, standing on the bedrock below and supporting 200-meter tower above, as shown in Fig. 3 – 4. This was a critical step to ensure the towers to remain upright. If they started to *tilt*, there could not be stopping trend. Before the builders could finish the bridge, they needed to solve another critical problem, it was *imperative* that *anchor piers* held the main cables were not able to move. But the ground here was so muddy and weak that the cables could pull the anchor blocks out of place. Overcoming this weakness would be the most difficult and dangerous job for the entire project.

Figure 3 – 4 The piles for a solid footing

At that time, China's Runyang Bridge would be the third longest suspension bridge in the world. Loads on the main cables would be enormous, the cables could drag the anchor piers at the force reaching 68 million kilograms. Designers needed to prevent anchor piers from moving or the bridge could collapse. To do this, builders would have to extend anchor piers at least 30 meters underground, which would be the largest anchor block that Chinese had ever built. But when engineers probed the river bank, they discovered that the soil was worse than muddy, it was full of underground water, this was a major *setback*. Engineers had to find a way to divert underground water, or muddy water would pour into any holes they were *dug*, and flowed so fast that could trap the workmen inside. The engineers devised a plan that they decided to block the groundwater inflow by erecting a gigantic underground dam. All around the 3400-square-meter site, workers dug a series of *trenches* (Figure 3 – 5), they filled the trenches with *reinforced concrete*. The walls kept the wa-

Figure 3 – 5 The trenches

ter out, making it safer to dig within. But there was still a problem that north shore workers were using the China's only *deep trench digger*. So on south shore, contractors were forced to take a risky approach——they decided to control the deep underground water by freezing the ground.

To do this, they needed to construct a huge *refrigeration plant*, which could chill the salt water to minus 20 degrees Celsius that salt water had a much lower *freezing point* than fresh water. Pipes circulated the water more than 30 meters down around the site, this system was like a *coil* in a large freezer as shown in Fig. 3-6. The ultra-cold water froze the soil into a meter thick wall, which could keep the water out. With groundwater no longer circulating in these two sites, the soil became safer to dig. The *excavation* went on 24 hours a day, as crews tried to finish it before the summer floods threatened their excavation site. These enormous *pits* were at least 30 meters deep, but every four meters workers needed to stop to *pour* concrete braces (Fig. 3-7). The deeper the workers went, the more dangerous the job became as groundwater was constantly pushing walls towards collapse. The deeper the hole was, the stronger the pressure would be.

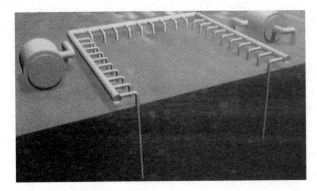

Figure 3-6 Pies around the dam

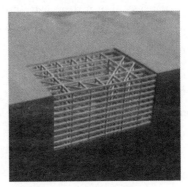

Figure 3-7 The concrete braces

Man Chung Tang: The deeper you dig, the more freezing you have to do, when the hole gets very deep, the water can come from the bottom, and can come very quickly.

Six months after the start of digging, the two anchor piers sank down at least nine storeys that looked like parking decks as shown in Fig. 3-8, and each anchor pier could park 7000 cars.

Figure 3-8 The finished piers were like car parks

Hundreds of men worked at the two vulnerable sites, and the danger was about to *escalate* that the *monsoon season* and the flooding were just weeks away. At that time, building the Chinese longest suspension bridge needed to construct two of country's largest anchor blocks, which needed to hold massive main cables securely, or the entire bridge would collapse. So far, workers had excavated 280,000 cubic meters of dirt, but engineers now concluded that it was too risky to continue to dig, they ordered crews to enter the next stage of the construction. Workers filled the two anchor piers with the *ballast* of 310,000 cubic meters of concrete and sand. The Runyang's engineers had won the first race against the rolling Yangtze.

3.7 The System of Main Cable Construction

Eleven months later, the region underwent the worst rain in five years. But by the time the flood did hit, work had moved on to the above ground part of the anchor blocks, where the cables were touched. But if they just put the end of the main cable in the concrete, it would pull out at once. So after each cable entered the anchor block, it was *dispersed* into 184 *bundles*. The cables spread from 1 meter to 9 meters wide, distributing the pulling force across 83 square meters. Each cable bundle was separately anchored in concrete, it was a forest of *steel wires* and it would be very vulnerable to *rust*. To protect it, the anchor block would be sealed entirely. The air here was constantly *dehumidified*, so rust couldn't form. From the anchor blocks, the cables climbed to the top of the bridge towers, and across the river. Each cable weighed over 20 million kilograms, too heavy to install straightway. The builders would have to string them one strap at a time. Firstly, they pulled a guide cable across the river by boat. To do this, they needed to block all ship traffic. They would later use a *steel rope* to pull more ropes across. It took 12 hours to get the rope up and over the towers. Once the cables were hung, they could open the river. Workers slowly built on the guide rope, they created a *cat walk* high above the water as shown in Fig. 3-9, which they used to string the cables, one bundle at a time. Each wire in steel bundles was made of ultra strong steel. It was so strong that just one could lift three cars. The contractor *pre-cut* every wire into a specific length. Every cable in a bundle has a different length, we have a lot of confidence in the calculation to the predict length.

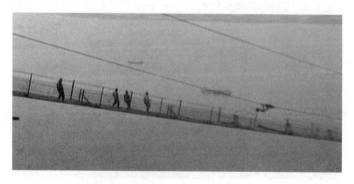

Figure 3-9 The cat Walk

Twenty-four hours a day, workers continued to install the Runyang's main cables. Into 7 months, they strung all 368 bundles, workers *compressed* bundles into two major cables, containing the 470,000 steel wires strung strongly end to end that would circle the earth three times. Workers *wrapped* each bundle of cables with an *interlocking* steel belt, then painted it to keep out moisture, The finished cables were almost a meter thick and 2.5 kilometers long, that would be the bridge's most expensive parts, and perhaps the most important.

Tom Ho (Structural Engineer): Whole weight of the whole bridge was supported by those main cables, because of the length of the suspension bridges, it could be a very large force.

The cable's greatest enemy would always be the *moisture* that caused it rust. To protect them, the builders installed *dehumidifiers* to pump dry air into the sleeve. Few bridges had this *sophisticated system*, it should keep the cables from rusting for 120 years.

China was building 40,000 kilometers of roads and bridges every year, which consumed a third of the world's steel production, and used half the world's concrete. This building boom may be the biggest in history. Almost 3 years after construction began on Runyang, workers were ready to enter the final stage, raising the road deck.

3.8 Key of Wind Resistance Technology

The design of these steel boxes was critical, because it could determine that the Runyang was survived for decades, or quickly fell apart. This happened to one suspension bridge in the United States. In July 1940. The Tacoma Narrows Bridge in Washington State opened to traffic, at that time, it was the third longest suspension bridge. Right from the beginning, it had a problem——light wind could even shake the bridge, it was quickly *nicknamed* as "Galloping Gertie". Four months later, the wind of just 67 kilometers per hour made the bridge twist and *undulate*, the bridge was shaking harder and harder, attaining to a *catastrophic* degree, called "flutter". After a few hours, the bridge could take no more. The Tacoma Narrows Bridge was doomed to collapse because the bridge deck was too flexible. It could still be standing today, if the designers strengthened the span with a *truss*, like Akashi Kaikyo Bridge in Japan.

Runyang Bridge was nearly twice as long as Tacoma Narrows Bridge, but engineers had also chosen deck design without trusses. Would flutter destroy the Runyang? The Runyang was designed to be China's longest suspension bridge, with the span of nearly 1.5 kilometers long. All of it was exposed to strong winds, and even typhoons. But builders were relied on *aerodynamic* design to keep it steady. The steel boxes that made up the bridge span were just 3 meters high, which had *slanted* edges like airplane wings, to allow wind to flow around them rather than push them into motion. In a giant wind tunnel, the builders conducted an experiment to *blast* the design in typhoon force skills as in Fig. 3-10. The deck heaved, but there was no destructive flutter.

Still the Runyang was at a critical phase, monsoon season was only five months away, and suspension bridge with deck uncompleted was the most vulnerable.

Figure 3-10 The wind tunnel experiment

Mark ketchum (Structural Engineer): *Girder* segments were almost *dangling* from the cable, if the winds blew too hard, they began to hit and may cause damage with each other.

The Runyang's builders faced yet another risk to the finish. They had special *gantries* to speed up work, the gantries moved along the bridge cables and hoisted each road section into place. Ninety days after lifting the first *slab*, the Runyang's work crew lifted the last one into position.

1490 meters' span of the third longest bridge at that time in the world was completed. Almost immediately, the tough job of maintenance began. Like the bridge cables, the span's 25,000 tons steel could *corrode*. But designers made it easy to inspect——a small electric car inside the decks made workers travel entirely in about 18 minutes. The bridge's *high-tech* control center was around, sensors captured every movement of Runyang. Runyang is in constant motion. This flexibility is unique to suspension bridges, it allows the immense structure to obey the nature's will, but will not break.

On 30 April, 2005, this suspension bridge opened to traffic, several months ahead of schedule. It took four and a half years to build, at a cost of 700 million US dollars. The region now had an economic lifeline between north and south. China is about to span the Yangtze in the 21st. In the next two decades, China is intending to build another 50 bridges above this river alone, China's most spectacular mega bridge may be still to come.

3.9 The Longest Sea-crossing Bridge in the World

Global Times: The Main Bridge of Hong Kong-Zhuhai-Macao Bridge runs through the whole line. The major construction work on the world's longest cross-sea bridge, which connects Zhuhai in Guangdong Province with Hong Kong and Macao, was completed on October 9, 2016.

A ceremony was held in Zhuhai to celebrate completion of the 55-kilometer cross-sea route and to mark the start of road surfacing and related work. "This means construction has entered its final stage." said Zhu Yongling, head of the administration bureau of the bridge. The bridge will cut travel time from Hong Kong to both Zhuhai and Macao from the current three hours by road or one hour by sea to a half hour drive, Zhu said.

Construction began in December 2009 at Zhuhai. The Y-shaped bridge starts from Lantau Island in Hong Kong with branches to Zhuhai and Macao. More than 400,000 tons of steel have been used for the 6.7km undersea tunnel and 22.9km bridge, enough to build 60 Eiffel Towers. British newspaper "The Guardian" named the mega structure as one of its "seven wonders of the modern world near completion". Before the project, the 36.48-kilometer-long bridge across the mouth of the Jiaozhou Bay in eastern Shandong Province was considered the world's longest cross-sea bridge.

Su Quanke, chief engineer with the administration bureau of the bridge, said the project has made history in many aspects. Located near the airports of Hong Kong and Macao and over one of the world's busiest shipping routes, where more than 4,000 vessels passing by every day, the project could not interfere with the daily operations of the vessels and the airplanes nearby. Maritime protection was also important as the construction region is overlapped with the reserve of Chinese white dolphins, an endangered species that enjoys top level state protection.

According to Su, this was the first time China has built an immersed tube tunnel in open waters. Two artificial islands made of steel pillars were created to connect the bridge, which was the first time this process has ever been used. Su said," The special position and the super-long status of the project were challenges. With a series of world advanced anti-erosion and earthquake-resistant measures, the bridge can be operational for 120 years."

Major parts, including bridge deck units, piers, girders and steel tubes, were produced at plants and then assembled piece by piece on the water with large floating cranes. The dolphin-shaped bridge tower weighs up to 2,786 tons. Two floating cranes, with a load capacity of 3,200 tons and 2,300 tons respectively, worked together to lift and roll the tower, setting a world record.

Su said, The 'China technology' and 'China standard' set by the Hong Kong-Zhuhai-Macao Bridge will influence the world market. If it were not for the progress of Made in China, we couldn't possibly have realized the construction in such a short time, considering the maximum load capacity of a floating crane was only 100 tons 15 years ago.

Words and Expressions

spectacular   adj. 壮观的
nightmare   n. 噩梦
submerge   v. 淹没
cable   n. 电缆，锚索
masterpiece   n. 杰作
construction   n. 建设，建筑物
daunting   adj. 使人畏缩的，令人胆怯的
clogged   adj. 阻塞的，堵住的

congestion   n. 拥挤，拥塞
suspend   v. 使悬浮
radical   adj. 激进的
unprecedented   adj. 空前的
collapse   v. 倒塌，瓦解
magnify   v. 放大，扩大
pile   n. 桩
segment   n. 片段，组件

crook	v. 使弯曲
specification	n. 规格
hoist	n. 起重机；v. 升起
scaffolding	n. 鹰架，脚手架
topple	v. 倾倒，倒塌
dismantle	v. 拆除，拆卸
precarious	adj. 危险的
vulnerable	adj. 易受攻击的，脆弱的
structure	n. 结构，建筑物
susceptible	adj. 易受影响的
weld	v. 焊接；n. 焊接点
cramped	adj. 狭窄的
ventilation	n. 通风设备
wrench	v. 扭伤，猛扭
joint	n. 接缝；v. 连接
snap	v. 突然折断
plunge	v. 投入，陷入
magnitude	n. 量级，震级
ordeal	n. 折磨，严酷的考验
destructive	adj. 破坏的，毁灭性的
pound	v. 连续重击，猛击
unscathed	adj. 未受伤的
seal	v. 密封
bolt	n. 螺栓，螺钉
rock	n. 岩石
string	n. 弦
flatten	v. 变平，使平坦
horizontal	adj. 水平的，地平线的
eliminate	v. 消除，排除
tension	n. 张力，拉力
obstacle	n. 障碍，干扰
float	v. 浮动，飘动
contractor	n. 承包商
showpiece	n. 展出品，展示品
ambitious	adj. 有野心的
slice	v. 切下，把…分成部分
watershed	n. 分水岭
ferry	n. 渡船；v. 摆渡
encounter	v. 遭遇，遇到
silt	n. 淤泥，泥沙
disastrous	adj. 灾难性的，损失惨重的
deluge	n. 洪水，泛滥
inundate	v. 淹没，泛滥
crumble	v. 崩溃，粉碎
pylon	n. 桥塔
deviate	v. 脱离，偏离
destroy	v. 破坏
navigate	v. 驾驶操纵，航行
slam	v. 砰地关上，猛力抨击
anchor	v. 固定，抛锚
pole	n. 支柱
storey	n. 楼层
tremendous	adj. 极大的，巨大的
bedrock	n. 基岩，根底
incredibly	adv. 难以置信地，非常地
footing	n. 基础，基脚
stilt	n. 高跷
tilt	v. 倾斜
imperative	adj. 必要的
setback	n. 挫折，退步
dig	v. 挖，掘（过去式 dug，过去分词 dug）
trench	n. 沟渠，壕沟
coil	v. 盘绕，把…卷成圈
excavation	v. 挖掘，发掘
pit	n. 矿井，深坑
pour	v. 灌注，倾泻
escalate	v. 逐步增强，逐步升高
ballast	n. 压舱物，压载物
unscathed	adj. 未受伤的
disperse	v. 分散，使散开
bundle	n. 束，捆
rust	v. 生锈，腐蚀
dehumidify	v. 除湿，使干燥
pre-cut	v. 预切，按规格裁切
compress	v. 压缩，压紧
wrap	v. 包，缠绕
interlocking	adj. 连锁的

moisture	n. 湿度，潮湿	welding torch	焊接炬
nickname	v. 给…取绰号	tied arch	拉杆拱
undulate	v. 波动，起伏	road deck	路面板
catastrophic	adj. 毁灭性的，灾难的	elevated highway	高架公路
truss	n. 构架，桁架	veer off	突然转向
aerodynamic	adj. 空气动力学的	run upon	撞上，偶遇
slanted	adj. 倾斜的，有倾向的	main cable	主缆索
blast	v. 猛攻；n. 冲击波	pulling force	拉力，牵引力
girder	n. 大梁，纵梁	anchor pier	锚墩
dangling	adj. 悬挂的，摇摆的	reinforced concrete	钢筋混凝土
gantry	n. 起重机架，龙门架	deep trench digger	深沟挖掘机
slab	n. 厚板，平板	refrigerating plant	制冷装置，冷冻厂
corrode	v. 侵蚀，腐蚀	freezing point	冰点，凝固点
high-tech	adj. 高科技的	monsoon season	季风季节
arch bridge	拱桥	steel wire	钢丝，钢绞线
suspension bridge	悬索桥	steel rope	钢索，钢丝绳
bridge span	桥跨	cat walk	猫道，角道
cable-stayed bridge	斜拉桥	sophisticated system	尖端系统
cable tower	索塔		

Activities——Discussion，Speaking & Writing

1. Listening Practice

Watch the video clips, write down the key words you have heard, retell the main idea in your own words and then check your writing against the Chinese scripts below.

Exercise 1：桥拱建造方式

Reference Script：首先工人在桥梁两端搭建临时桥塔，高度高达120多米，他们从桥塔用钢缆系着桥拱，每个拱段由两根缆索支撑，桥拱一块一块朝中间搭接。

Exercise 2：土壤无法承受桥拱重量

Reference Script：但黄浦江的土壤太松软，无法承受卢浦桥的重量，直接导致两端会向外滑，这样滑下去桥拱终将倒塌，除非工程师把两端连结在一起。

Exercise 3：拉杆拱原理

Reference Script：这叫做拉杆拱，造桥者在桥拱底座间拉起16条缆索。桥梁完工后还能看见8条，另外8条会藏在桥跨内部。这些缆索联合起来就像是射弓的弓弦，少了它们的拉力，桥拱会变平。

Exercise 4：桩柱支撑桥塔

Reference Script：但长江的淤泥特别厚，为了获得坚固的基脚，造桥者必须在每个桥塔基地沉陷 32 根桩柱。桩柱就像巨大的高跷，竖立在下面的岩床上，支撑上面 200m 高的桥塔。

Exercise 5：冻结墙壁

Reference Script：管道将水循环至 30 多米以下，环绕整个场地，这个系统就像大冷冻库里的线圈。超冷的水把土壤冻结成 1m 厚的墙壁，冻土墙壁把水阻挡在外，这样地下水不会循环至这两块建筑场地，挖掘土壤的工作会变得比较安全。

Exercise 6：尖端防锈系统

Reference Script：造成锈蚀的湿气是缆索的天敌，为了保证缆索不被锈蚀，造桥者装设除湿机把干空气注入缆索套里。很少有桥采用这种尖端系统，它能保证缆索 120 年都不会被锈蚀。

Exercise 7：桥梁抗风原理

Reference Script：组成桥跨的钢箱梁只有 3m 高，钢箱梁边缘是倾斜的，就像飞机的机翼，可以让风绕着边缘流过去而不会迫使桥梁摇晃。风洞实验时，在巨大的风洞里，造桥者按照台风的风力对这个设计施予风荷载，桥面上下起伏，但没有发生毁灭性的颤振。

2. Group-work：Presentation

Group (5 to 7 members) discuss and give your opinions about **the world's longest cross-sea bridge**. Choose one theme below for you presentation. Clearly deliver your points of the following themes to audiences. Each group collects information, searches for literature, pictures, etc. to support your points. Prepare some PowerPoint slides.

10 minutes per group (each member should cover your part at least one or two minutes).

NEED practice (individually and together)!! **Gesture** and **eye contact. Smile** is always Key!!

Themes on the Hong Kong-Zhuhai-Macao Bridge

(1) Construction Course and Bridge Location
建设历程和桥梁位置
(2) Architectural Structure and Design Parameters
建筑结构和设计参数
(3) Equipment and Facilities
设备和设施
(4) Management Institutions and Traffic Flow
管理机构和交通流量

(5) Construction Achievements and Technical Challenges
建设成果和技术难题
(6) Study on Durability, Wind Resistance and Seismic Resistance of Condyle Bridges
桥梁耐久性和抗风、抗震研究
(7) Innovative Application of New Technology, Equipment and Materials
新技术、新设备、新材料创新应用
(8) Cultural Characteristics and Value Significance
文化特色和价值意义
(9) Application and Practice of BIM Technology Innovation
BIM 技术创新应用与实践
(10) Application of Intelligent Health Monitoring System
智能健康监测系统应用

3. Independent work: Writing

Search for **the Hong Kong-Zhuhai-Macao Bridge** in relevant Index, like Engineering Index(EI), Scientific Citation Index(SCI) and China National Knowledge Infrastructure(CNKI). Then write a report independently on **the "Green" design and technology in the world's longest cross-sea bridge**.

Unit 4
German Highway

Teaching Guidance

Read articles, pay attention to the **Words and Expressions** and related **sentences** and **paragraphs**.

This road is built for speed. It realizes people's dream of running at a high speed. This is the *Autobahn*. The combination of unlimited speed and German *precision engineering* has attracted drivers from all over the world. We will introduce the function of Autobahn, derivative problems and the way to deal with highway driving problems.

4.1 Unlimited Speed & Precision Engineering

On the road, it is a modern classic. Human has dreamt of a *barrier-free* road. When speed is increasing, the dream is even more exciting. It is charming and dangerous to drive on the road with full *throttle*. The German *Autobahn*, the word is generally translated as motorway, is one of the world's most technologically advanced highway systems. But for those who love the road, the first word comes to mind is speed not engineering.

In pursuit of this thrill, Kurt Lotterschmid built his highway car entirely by hand, and he estimated that he spent over a million Euros.

Kurt Lotterschmid: I have a dream. I always want to build a car completely by myself.

This car is called "low-tech" car, however, it is fast. Kurt made a *speedometer* for it.

Kurt Lotterschmid: It has a top speed of 400 km/h, but the tyres are only capable of 380 km/h. But I have the power to go for with it under 400km/h.

4.2 The Safest Highway in the World

However, before people begin to think that German Autobahn has turned citizens into speed demons, there are people opposed to high speed driving. Peter Seywald is one of those opponents. He is the President of Regensburg VCD, that is the Traffic Club of Germany.

Peter Seywald: There are about 6,000 people a year died in car accidents in Germany. Officials think this is much better than in the 1970s, when about 20,000 people a year died in car accidents. But I was shocked that it is said 6,000 deaths a year is progress.

Even for student driver Silke Urmetzer thinks the excitement of learning to drive fast on this super road is exciting.

Silke Urmetzer: When you are driving slowly, it is easy to control. While you drive so fast, and you make a mistake, you don't have your car on control.

Being out of control is the horrific flip side of driving at super speed. Losing control and driving too fast is like driving in the rain. When you are driving at speed more than 240 kilometers per hour, a tiny swerve or one slip *distraction* can lead to devastating outcomes.

The average person's reaction time is one and a half seconds. If you're driving at 100 kilometers per hour and you slam on the *brakes*, you still travel 120 meters before you come to a complete stop. If you slam on the brakes at 300 kilometers per hour, you're still going almost one kilometer before you stop.

Roland Benning (Helicopter pilot): German highway accidents are always fatal. *Airbags* don't work when you're driving at high speed against a wall, and they don't work when you're going over 200 kilometers per hour.

German Autobahn is prone to fatal speeding accidents, but statistically it is one of the safest roads in the world, with a lower death rate per kilometer than the United States. Why? Germany has equipped this fascinating and dangerous highway with innovations, facilities and staff. Road crews also have the first aid skills of doctors and the repair skills of car *mechanics*. The technology on the highway, from the gleaming command center to the high-tech sign-washer on the ground floor, is part of the world's most famous highway backup system. The road is so important that the details from road *construction* to letter size on the traffic signs have been carefully planned.

4.3 Design and Maintenance for Autobahn

Dr. Tomas Linder comes from the department of Operations and Traffic for the Bureau of Highway Authority of South Bavaria.

Tomas Linder: Germany's road network is designed to connect major cities and important areas in the country, with extremely high standards.

Fig. 4-1 shows the simulation image of Autobahn. There are several reasons that Autobahn can outperform the world. Firstly, the Autobahn is thicker than most highways. Depending on climate and local ground conditions, the thickness of roads ranges from 55cm to 85cm in depth. Even the aeroplane 747 lands on it, the road will not sink one centimeter. Therefore it has a foundation solid enough to cope with *pounds* day and night. Another factor that helps make this road supreme is one of its unique features. It is also one of the main reasons that you can safely drive so fast on it——it is the gradient of the road.

Figure 4 – 1 The simulation image of Autobahn

Tomas Linder: We have many design rules for the highways, containing the grading, the curves, minimum radius and so on. Things like those enable the highway to be driven with high speeds.

The idea is to replace deep hills with *gentle slopes* to accomodate this high speed. After all, you can go faster as the road is flatter. Its maximum slope is of 4%. If you are driving on a 20% grade, you can climb 200 meters in one kilometer, as shown in Fig. 4-2. While on a 4% grade, you are nearly to drive 5 times distance to make the same climb. Maintaining a 4% incline seems easy, but when considering German *undulating terrain*, extending all the way from *Baltic Sea* to *Alps*, you will find the *toughness* of this task. It is sound to say that German crews had to literally" move mountains" to build the Autobahn.

Figure 4 – 2　The 20% slope

The final reason why the Autobahn is super awesome in the world is super wire, which is shown in Fig. 4-3. Overlaid on this continuous highway is a system of *sensors*, cameras, electronic signs, communication points and traffic computers that continuously monitor every aspect of the Autobahn. Some sensors are hard-wired into the system, while others are *solar-powered* and *wireless*. They feed back everything, from temperatures and weather conditions to traffic and accidents. Whatever you do on this highway, every action on this road is closely monitored. Of course, this German project does have high price. to keep drivers' happy driving, Germany spends over 460,000 Euros *anually*, supporting the *maintaining* of the road, which is more than

Figure 4 – 3　Super wire systems

twice the cost of the American interstate systems every kilometer.

Tomas Linder: Speed costs money. Our road system is excellent. It is well maintained, which contains technical equipment. Drivers can be happy in our Autobahn.

4.4 Perfect Monitoring System

Indeed, there are some occasional speed *traps* on Autobahn. Only about two-thirds of the 12,000 kilometers of Autobahn have no speed limit, which is still impressive for speed freak. However, German police in Autobahn fight speed with speed, and they have got cooler police cars than most of the people they chase. Their fleets of cop cars are the car collectors' dream team, including Mercedes, BMW and Audi, the *horsepower* of which is able to match racing cars. The vehicles are *outfitted* with standard police equipment, such as radio, computers, lights and camera, but there is mystery under the hood. Besides having an extra set equipment to strengthen *shock absorber*, it is also one of the few vehicles that does not need a 250km speed limiter. One reason German police cars need extra horsepower is that a family in **Pfeffen** Hausen whose business is to make cars that push the limit of speed. Many drivers believe that the perfect car for Autobahn is the Ruf Porsche.

The Ruf family has been in car business since 1939 and developed from a simply maintenance corporation to a full customization of Porsche. In 1981, Alois and Estonia Ruf began *manufacturing* their own cars based on Porsche chassis. Their hand-made cars are considered as the most advanced in the world.

Alois and Estonia Ruf: We are in the only country in the world that we have freedom of driving in Autobahn. Our cars are made to *outperform* any other cars in the world. German cars were always the best cars in the world. We, the Ruf family, will take the best car, namely the Porsche, to turn to another better level.

The R Turbo, is the top car of Ruf family, whose chassis is from Porsche 996 turbo and other parts are redesigned. It is a machine for German motorway enthusiasts. The four-wheel-drive R turbine can accelerate to 100kph in less than four seconds, with a top speed of about 350km/h. How fast is it? The taking off speed of a large commercial jet is about 290 kilometers per hour. So how does the *patrolman* in Autobahn battle this kind of speed? He relies on time-tested tools.

First is the technique, Autobahn police tend to block in front of the target vehicle, which can prevent routine traffic stop from becoming high speed chases. Then it is technology. Many commercial vehicles have *telegraphic speed recorders* mounted in site. The police can check out the speed history and *issue a citation*. However, sometimes the traditional methods are not enough to this extreme road, and extreme measures are needed to slow cars down. How do German traffic police train for super speeders? The answer is that they can play video games, as shown in Fig. 4-4. This is a specialized police *driving simulator*, located in police training center in **Sulzbach**, about 200km north of Munich. It might be the only high speed simulator of this kind in the world, which is built like aircraft flight simulator. This equipment is simulating the police BMW. This simulator has a huge panoramic screen. To be realistic, there are images in the rearview mirror. The simulator can replicate all kinds of driving situations, from small towns, *country roads*, to urban *congestion* and high-speed runs on Autobahn. For *trainees*, understanding what people will react is crucial. During a superfast chase, sirens are *ineffective* because drivers going top speed can't hear them. By the time the driver sees the blue flash light in his *rearview mirror*, the police are already about to overtake him. Extensive simulator training hones the reaction time to split seconds and forges vechicle skills far beyond those of average drivers.

Figure 4 – 4 Driving simulators for police cars

The second line of defense for the police is the locomotive police and the super locomotive. One type of bike is BMW K1200GT with a top speed of 240 kilometers per hour. It can accelarate the speed from 0 to 240 kilometers per hour in just 3.7 seconds. Motorcycling on Autobahn is a dangerous job. Most drivers on the road have seat belts, air bags and rolling chassis to protect them, while motorcycle police only rely on his helmet, leather coat and his skills. Sometimes those are not enough on the Autobahn, so the police must have another stealthier tool, high-tech *surveillance* that can ensure the Autobahn drivers to follow the rules, even when the police are not on site. This method is called photo enforcement, but it might be more likely to be called as highway watcher. Undercover vehicles park alongside the roads, as well as signs and underpasses are equipped with cameras, which can take pictures of speeding cars. The *negatives* are processed scanning into the computer system that digitally enhances the picture to create a clear image. After that, it scans the driver's face of the offending vehicle and intensifies that image. Then, the two pictures put together along with the citation are sent to the driver who thought he has got away with it.

Photo enforcement works because it's invisible. We all know that speeding drivers will slow down when they see a police car. And drivers usually take any chances that they accelerate as soon as they pass the police car. This time, the *undercover petrol* comes, using high-tech equipment, such as *concealed* video cameras and laser speed sensors. These officers shuttle through traffic flow, scouting law breakers on Autobahn. Once they have identified the speeder, they top their car with a blue light and try to pull alongside the offender. Assuring of the eye contact, they wave a stop signal which instructed the driver to follow them or pull over. Then the police issue a citation to the drivers.

Officer Wolfgang Steiner (Erding Highway Patrol): The fastest car that I had followed was going at 245 kilometers per hour on a zone with 120 kilometers per hour limit. The fine for this kind violation is 375 Euros. The drivers would also get his or her licenses suspend for 3 months.

The work of undercover police officers is exciting, and sometimes the excitement on Autobahn can be thrilling, always dangerous.

Officer Christian Renftle (Erding Highway Patrol): Whenever we are driving with high speed on the Autobahn, it is really dangerous. When we chase a car, other drivers don't recognize us as the police. They sometimes get in the way and slow us down, which is really dangerous.

Even though the Autobahn is one of the fastest roads on earth, under the close *supervision* of police force, it remains one of the safest roads in the world.

4.5 Efficient Road Rescue Team

Rescuer helicopter pilot Roland Benning is doing routine maintenance, making sure his helicopter is filled up and ready for the next call.

Roland Benning: It looked easy at this moment, but in one second, situations may change to a 180 degrees turn.

At this moment, a speeding car lost control and this pilot is about to take the 180-degree turn he mentioned. Benning is a pilot of the Allgemeiner Deutscher Automobil-Club (a club for German automobile), also known as ADAC. This club is the Autobahn's lifeline. ADAC is a powerful automobile association, with a scope of responsibilities ranging from tire fixing to *aerial* rescue. This road rescue team goes so far beyond its responsibility that it is nicknamed as "yellow angels" (Because the body of the car is yellow as shown in Fig. 4-5). The ADAC operates a fleet of 1700 rolling repair shops as well as 36 helicopter bases throught out the country. Martin Hauser is the team leader of the ADAC roadside assistance center in Landsberg, just beside Munich.

Figure 4-5 The road rescue car ADAC

Martin Hauser: If somebody has broken down on the Autobahn, there is a special number to call. When we receive the call, we will ask some questions like location, car type, break-down situations. Also we will ask if the vehicle needed to be towed or to be repaired at the location. When typing this information in the computer, the dispatcher will receive and *dispatch* the crew who is close to the location to take the call.

When the ADAC comes to the rescue, it means more than towing. "Yellow Angels" fix the vehicle on the spot. More than 80% of the vehicles can be repaired on site. In auto mechanic's car, it is full of repair tools, which is of convenience for "Yellow Angels" to *drill* and *weld* at the shoulder of the road. Each of the "Yellow Angel" cars is *custom-made*. The custom king for ADAC is **Alois** Stiegler, who is ADAC's main *bespoke dealer* and has been turning garages of vans for 25 years.

Alois Stiegler: Per year, we have converted about 200 to 250 vehicles from regular manufacturers into our own rolling car repair shops. Each car needs about 55 hours until it completely matches our needs.

These converted "Yellow Angel" vans contain a wide sort of equipment, from basic *wrenches* to advanced electronic detection tools. The devices have enabled 83 percent of the cars that broke down to be repaired and to *speed along* on the roads again.

4.6 Efficient Helicopter Casualty Rescue

There is another difference between the ADAC and any other auto club in the world——These "Angels" aren't merely on the ground, they can fly as well. Given the high speed driving on the Autobahn, when crashes do happen, they can be disastrous. To deal with these horrific accidents, 36 ADAC helicopters are located across Germany, allowing them to respond to any location of the Autobahn within 15 minutes or less, to extract victims from the scene and transport them to the nearest *trauma center*, as shown in Fig. 4-6. The system developed for Autobahn has proven to be so effective in saving lives that this trauma rescue has been replicated worldwide. As soon as the call rings in the ADAC command center, Roland Benning will respond quickly.

Figure 4-6 The helicopter rescue of ADAC

Roland Benning: Once the beeper rings, I will run directly to the helicopter and start the engine. The assistant will make a call to the rescue center, inquire the information where the accident happens and tell me where to fly. The whole preparation process is less than 2 minutes.

Together with the Roland Benning is the doctor Angelica Grunes. This time, they spot the wreck and it is on the *artery* road near Autobahn. This accident is very awful.

Angelica Grunes: When we got there, we saw a car crashed into a tree. It was badly damaged.

This accident involved three passengers. Two people died at the scene and one in critical condition was taken to the nearby hospital by the ADAC.

Angelica Grunes: Probably this accident was due to fast driving because there was no other car involved. In my opinion there should be a speed limit because many of the accidents are young people, they're about 18 years old and they've just gotten their *driver license*. These young people may not be able to control the high speed.

Roland Benning: If you see something terrible, sometimes like a broken hand or a broken arm, or a dying child in your arms, you should talk to someone after the mission. After that you may feel better that day.

This is the only way that can make sure Benning to continue to fly tomorrow.

Roland Benning: You should tell someone. Only by that, can you sleep soundly and face the next task with ease.

4.7 Autobahn Control Center

Looking back, we find that the Autobahn even has a *pedigreed* speedy, which was formerly a race track. Construction started in 1913 on an experimental roadway. This 19-kilometer highway was built in south of Germany and was initially used as a speedway.

Dr. Cornelia Vagt-Beck (Scientific Director, German Road Museum): The Autobahn system has been expanded greatly since 1930s. That happened because people realized it was much easier to drive on a broad and smooth road of the Autobahn, which allowed for higher speed.

By 1942, the Autobahn had been built more than 3,900 kilometers, connecting most major cities in Germany. Eight years after the end of World War II, construction of the Autobahn *resumed*, and this monstrous road has been growing ever since. But the road is not just continuous thick concrete, this road practically "thinks of itself". The cutting-edge infrastructure behind this highway is as *compelling* as its bleeding driving speed.

One of the nerve centers of Autobahn is located at Munich, which is called the command center. This center is directly responsible for over 1200 kilometers' Autobahn, which are the busiest road segments in Germany. One of the chief jobs of this center is to prevent traffic jam.

According to a staff in this command center, there are 45,000 cars per day on Bavarian roads, and in Munich, there are 140,000 cars and 180,000 on holidays. These centers control

Figure 4-7 A sensor controlled by computers

the congestion with their eyes, which means thousands of cameras monitor every kilometer of this super highway. Where the visual check is not possible, sensors can electronically feel what is going on for the road. Technology enables Autobahn to hear sounds, detect motion and sense heat. It might be the smartest road in the world. All that data is sent to the computer in the control room, and the computer will process and transfer that information. With all the electronic sensors continuously monitoring the highway, you can imagine how large the volume of data is that will flow into command center's computers. Here is how sensors work to control the traffic jam: Sensors detect the signs of traffic jam outside Munich, maybe sensors find the flow slow down or maybe rada speeder detects cars barely creeping along. One computer in command center receives the numbers and relays a set of instructions to the electronic road signs in the slow traffic areas. Suddenly, symbols in the road signs change and the *shoulder line* becomes an extra lane, in order for cars to move on at their usual speed. Figure 4-7 shows the image of a sensor. All is fulfilled under computer control. In addition to changing lanes, there is a system for open road shoulders, which is a simple but effective way to relief traffic. Under peak traffic loads, the road shoulder is remotely converted into an extra lane, providing additional room for vehicles to keep moving as shown in Fig. 4-8. The only time that a human comes into play in this system is for manual lane conversion when an accident happens or constructions are conducted or by the police's instruction.

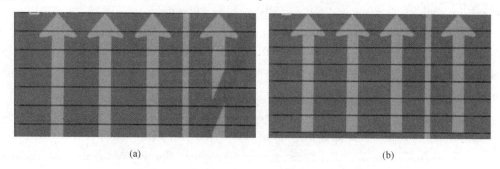

(a)　　　　　　　　　　　　　　　(b)

Figure 4-8 The road shoulder conversion for rush hour
(a) No passing on road shoulder; (b) Road shoulder opens to relief traffic

Tomas Linder: The police will notify us if they need to change signs. They will give a telephone call to the command center. And then the operator will make a change on the signs as the police instructed.

4.8 Meticulous Maintenance & Well-designed Highway

While the command center is the brain behind the scenes of this highway, maintenance operation is out there every day, dealing with the physical stuff, keeping this road glossy and beautiful. Maintenance is what Autobahn earns its glowing reputation. Every day thousands of specialized vehicles are crossing Germany, working round the clock to make sure the road ways are safe and in perfect driving conditions.

Tomas Linder: The maintenance means shoveling the snow in winter and weeding the grass in summer, it also means repairing the road signs and safety barriers. Once there is an accident, the maintenance personnel will go out and assist the repair work.

Klaus Seuferling is the director of the Munich maintenance center and he is very clear about his mission.

Klaus Seuferling: Our first priority is to ensure the safety of the traffic and keep the traffic flow smoothly.

The biggest part of dealing with maintenance work is to deal what's on the road, which are the traveling fast vehicles. Workers rely on specialized vehicles to keep them safe.

Klaus Seuferling: For people working with that equipment, it is very dangerous. Not just for them, but for the drivers on the Autobahn. To protect themselves, they need special cars. It is a kind of car that has only one purpose——To keep from other vehicles. It enables cars to avoid lands under construction and it is equipped with special energy absorbing materials. This material can absorb the impact of high speed *collision* and prevent the workers from being seriously injured.

The repairmen also have a number of other specialized vehicles, such as the **Unimoc**, which is made of a Mercedes headstock and diverse attachments, as shown in Fig. 4-9. This car can do a lot of work. The **Unimoc**'s speciality is washing road signs.

Figure 4-9   Unimoc for road maintenance

Though there are lots of expensive machines to do their jobs, the best part for maintenance is the patrol man.

Klaus Seuferling: The task of every Autobahn crew is to check road conditions that they are responsible for. So every day they drive through the Autobahn to see what's wrong and what might happen. And they look for precise spots to be repaired.

The final factor that makes the Autobahn suitable for driving fast is the construction. The road is built solidly, most construction is new, which does not need much maintenance, because it hardly ever breaks down, with few cracks or *potholes*. One of the reasons that the road has no potholes is that it is supersized. The average thickness of concrete on the Autobahn is 70 centimeters, comparing that the average thickness of American highways are only about half of it. Concrete expands when it is hot, and contracts when it is cold. The thinner the road surface is, the more easily to be cracked by the environment it will be.

Another reason that Autobahn doesn't crack is that its construction allows it to shed water fast. Surprisingly, water is the main cause to damage the highways. Rain does more harm to this road than millions of tires traveling over 200 kilometers per hour. Engineers prevent rain damage by tilting the entire Autobahn. Here's how it works: The road is sloped 2.5% to one side, which enables water to drain to that side. After filtering through a layer of *porous* concrete, the water flows to a net of pipes and finally to the *reservoir*. The rainwater is stored there, protecting the environment from the pollutants that water contains.

Cutting-edge technology, innovative engineering, and meticulous maintenance all make the Autobahn dynamic and reactive. You don't just drive on this road, you drive with it.

A bus driver: Overall, it is a well-designed road. The quality of the road is great. You can often see maintenance engineering. The road is well maintained and comfortable to drive. It's *definitely* a well-designed highway.

4.9 Top One Driving Technology in the World

Germans love their cars. But in Germany, if you want to drive fast, you have to pay a considerable price. To drive fast is a *privilege* after a lot of hard work. Every driver in Germany must undergo special training before they get their driving license. Earning the right to drive on the Autobahn or any other highways takes about two months, involving intensive study for tests. It would also spend a lot of money for a driver license. The cost for a driving school is about 1500 Euros. Because of this rigorous training, it can be assumed that German driving should be world-class. When students drive on the highway for the first time, they are naturally a little scared.

Florian Ulleweit (Driving Instructor): At first, they don't want to drive very fast, because they have only driven in the city. Some people have driven on country roads only once or twice. They will be scared, but after several times' training, they want to go fast.

Driving instructor Florian Ulleweit has taught thousands of students how to master this highway. He has found that the biggest problem for students on the highway is that their bodies become physically used to high rates of the speed, and when they exit the highway, they don't realize how fast they are going.

Silke Urmetzer (Student Driver): My teacher told me two things, you have to always look in the rearview mirror when overtaking. And the other thing is that you can speed up when you're on the range with others. After that you can drive safely on the highway.

4.10　The Unity between Drivers & Cars

Because of the Autobahn, Germany has been one of the few places in the world where the cars and the drivers can be connected.

Alois Ruf: How can I describe it? It is a feeling of a unity of cars and drivers. You and your cars become one.

The couple——Alois and Estonia Ruf, has been helping drivers become one with their cars for more than 20 years. Watching them travel on this road with over 250 kilometers per hour, it is obvious that they have a special relationship with the roads.

Estonia Ruf: We just want to achieve what the guests want to achieve, which is to give full play to the horsepower and speed of the car. I watched the car speed all the way from zero to 330 km/h, we all feel very happy. The car also gave full play to its performance, the so-called performance is not a blind pursuit of speed, but to be able to stop at will.

Estonia helped to design a 75,000-Euro entry-level car called Ruf 3600S. For her, these custom-built cars aren't just a means of transportation, they're works of art.

Estonia Ruf: A car is like an *orchestra*. As long as one part is out of tune, the rest cannot perform. For example, one violin is out of key, the whole concert will fail. This is the same with a car. We hope to find its defects. We usually think is it performing well? Are the brakes working properly? Can we stop at will? Is the transfer force strong enough? Is the shock absorber moderate? Will the car jump too severely? We tried to figure out these questions all the time and I have learnt a lot about them.

Alois Ruf: When Estonia sits on the copilot seat, she can feel something wrong with the car.

A sense of pride underlies the Autobahn. At first glance, it looks like a series of roads and freeways, no different with other highways. But after peeling back the layers to see what really makes the Autobahn, an unstoppable praise for precision and *expertise* comes to light. From meticulous maintenance crews, to the innovative operations to connect and control, all systems equipment, technology and people come together to cooperate well. If you look just a bit deeper, all of those ingredients merge together and go down one road——The road that leads freedom to drive fast.

Words and Expressions

Autobahn　n. 德国高速公路　　　　　throttle　n. 马力
barrier-free　adj. 无障碍的　　　　　enthusiast　n. 狂热者

speedometer   n. 速度计
distraction   n. 分心
brake   n. 刹车系统
airbag   n. 安全气囊（汽车）
mechanics   n. 力学
construction   n. 建造
pound   n. 重击，敲击
Alps   n. 阿尔卑斯山
toughness   n. 韧性，难度
sensor   n. 传感器
solar-powered   adj. 太阳能的
wireless   adj. 无线的，无线电的
anually   adv. 每年地
maintain   v. 维修，维护
trap   n. 陷阱
horsepower   n. 马力
outfit   v. 配备，供应
manufacture   v. 制造，生产
outperform   v. 胜过，比…做得好
patrolmen   n. 巡逻
congestion   n. 堵塞
trainee   n. 受训人员
ineffective   adj. 无效的
surveillan ce   n. 监督，监视
negative   n. 底片
conceal   v. 隐藏
supervision   n. 监督
aerial   adj. 空中的
dispatch   v. 发送，派遣
drill   v. 钻孔
weld   v. 焊接

custom-made   adj. 定制的
wrench   n. 扳手
artery   n. 干道，主流
pedigreed   adj. 有血统的，有来历的
resume   v. 重新开始，继续
compelling   adj 引人注目的，令人信服的
collision   n. 碰撞，冲突
pothole   n. 凹坑
porous   adj. 渗透的
reservoir   n. 蓄水池
definitely   adv. 明确地，肯定地
privilege   n. 特权，优待
orchestra   n 管弦乐团
expertise   n. 专门技术，专业性
precision engineering   精准的工程学
monitoring system   监控系统
gentle slope   缓坡
undulating terrain   起伏的地形
Baltic Sea   波罗的海
shock absorber   减震器
telegraphic speed recorder   电子速度记录仪
issue a citation   开罚单
driving simulator   驾驶模拟器
country road   乡村道路
rearview mirror   后视镜
undercover patrol   便衣警察
bespoke dealer   订销商
speed along   高速前行
trauma center   外伤中心
driver license   驾照
shoulder line   路肩

Activities——Discussion, Speaking & Writing

1. Listening Practice

Watch the video clips, write down the key words you have heard, retell the main idea in your own words, and then check your writing against the Chinese scripts below.

Exercise 1：道路厚度设计

Reference Script：德国高速公路能傲视全球有几个原因，其一是这条路比大多数公路都厚，依据天气和当地地面状况的不同，它的厚度介于 55~85cm，就算一架 747 飞机降落在上面，路面也不会下降 1cm，这几乎是美国洲际公路平均厚度的两倍。因此它具有坚固的基础，足以应付夜以继日的碾压。

Exercise 2：缓坡设计
Reference Script：工程人员用缓坡取代陡峭的山坡来提升行车的速度，因为路面越接近水平，车速就会越快。这条公路的最大坡度为 4%。坡度对驾驶的影响是，你在 20% 的斜坡上行驶时，你会在 1km 内爬升 200m，在 4% 的斜坡上，你要开上 5 倍的距离才能爬升 200m。保持 4% 的坡度好像很容易，但是当考虑道路会经过德国起伏的地形，从波罗的海延伸到阿尔卑斯山脉时，你就会发现这其实并不容易。

Exercise 3：完美监控系统
Reference Script：德国高速公路能称霸全球的最后一个原因就是它有完美的电子系统，这条绵延不断的公路上铺满了侦测器、摄影机、电子路标和通信点。而电脑也不断地监视着各种状况，有些侦测器是有线连接的，有些则是太阳能无线侦测器，它们会传回温度、天气、车流量和意外事故等资料，任何在这条公路上的举动都会受到严密监视。

Exercise 4：高速驾驶模拟器
Reference Script：这是警车专用的驾驶模拟器，它位于稣兹巴克的警察训练中心，稣兹巴克位于慕尼黑北方约 200km，这种高速驾驶模拟器可能是绝无仅有的，它就像飞机飞行模拟器，只是它模拟的是警用宝马，这个模拟器有个巨大的全景屏幕，为求逼真，它的后视镜也有影像。

Exercise 5：ADAC 道路救援组
Reference Script：ADAC 是个强大的汽车协会，负责补胎、空中救援等任务。这个道路救援组经常胜任他们职责之外的工作，因此被誉为"黄天使"，ADAC 旗下有 1700 个运行维护站，并在德国各地有 36 个直升机救援基地。

Exercise 6：直升机伤患救援
Reference Script：为了处理这些严重车祸，ADAC 在德国各地部署了 36 架直升机，让他们能在 15 分钟内到达高速公路的任何地点，将伤者运离现场，通往最近的外伤中心。这套为德国高速公路发展出的系统，救援成效非常卓著，因此全球各地纷纷仿效德国的伤患救援。尽管有世上最好的系统，ADAC 的直升机救难员仍无法预防车祸的发生。

Exercise 7：道路控制中心预防"大塞车"
Reference Script：这些数据资料就成了高速公路的"直觉"，它的运作方式如下：侦测器在慕尼黑外发现塞车的微兆，也许是摄影机发现车速变慢了，也许是雷达测速器发现车辆几乎没有前进，控制中心的某部电脑收到资料后，便会对塞车路段的路标传送指令。电子路标的符号会突然改变，路肩成了新增的车道，车辆又开始以高速行进，这一切都是电脑控制的。

2. Group-work: Presentation

Group (5 to 7 members) discuss and give your opinions about **the world's intelligent transportation**. Choose one theme below for your presentation. Clearly deliver your points of the following themes to audiences. Each group collects information, searches for literature, pictures, etc. to support your points. Prepare some PowerPoint slides.

10 minutes per group (each member should cover your part at least one or two minutes).

NEED practice (individually and together)!! **Gesture** and **eye contact**. **Smile** is always Key!!

Themes (Major topics included but not limited to)

(1) Transportation Planning and System Optimization
交通运输规划与系统优化
(2) Theory and Application Technology of Intelligent Transportation System
智能交通系统理论与应用技术
(3) Traffic Control and Information Technology
交通控制与信息技术
(4) Transportation and Socio-economic Development
交通运输与社会经济发展
(5) Environmental Protection and Sustainable Development
环境保护与可持续发展
(6) Green Transportation and Low Carbon Transportation
绿色交通和低碳交通
(7) Innovative Application of New Technology, Equipment and Materials
新技术、新设备、新材料创新应用
(8) Application and Practice of BIM Technology Innovation in Transportation Infrastructure
交通基础设施 BIM 技术创新应用与实践
(9) Application of Intelligent Health Monitoring System for Roads and Bridges
路桥工程智能健康监测系统应用

3. Independent work: Writing

Search for **the Intelligent Transportation** in relevant Index, like Engineering Index (EI), Scientific Citation Index (SCI) and China National Knowledge Infrastructure (CNKI). Then write a report independently on **Green Transportation and Low Carbon Transportation.**

Unit 5
Beijing-Shanghai High-speed Railway

Teaching Guidance

Read the article, pay attention to the **Words and Expressions** and related **sentences** and **paragraphs.**

The architectural wonder, Beijing-Shanghai High-speed *railway*, regarded as the most ambitious high-speed project started its *construction* in 2010. Chinese two greatest cities, Beijing and Shanghai have been linked by the fastest transport, and the line has broken the world record for the longest high-speed track in the world in 2010.

It was the first time that anybody constructed 1,318 kilometers of railway in 2010. Not only that, but at an *unprecedented* speed of 380 km/h, the new line has hold the fastest passenger train on wheels, pushing the limits of railway engineering, making the Beijing to Shanghai line the fastest high-speed railway in the world in 2010.

5.1 The Longest High-speed Railway in the World in 2010

Beijing is the political heart of China, the Olympic city and the country's cultural capital. Shanghai, the world's most *populous* city, home to the 2010 EXPO, is China's thriving financial center. Before the new line, traveling between the two great cities involved a ten-hour-line built a century ago. But that was about to change. Construction has begun on a new high-speed rail line that has completed the journey under four hours. The super-high-speed line can not only stimulate the economies of the two cities, but also connect a quarter of the Chinese population living along the railway. But to break the four-hour *barrier*, engineers had to build trains capable of carrying passengers at 380 kilometers per hour, a new world record for travel on trains. To avoid disaster, every part of the railway, from the train to the *track*, bridges, signals and stations, would have to be designed from scratch. For the designers and engineers, the new line was a truly unknown journey.

5.2 Trains Body Design Competition

The chief designer of the record-breaking project was in charge of the biggest responsibility to build the train. The design team had put a lot of attention to the design of the train, they wanted the design to be technically brilliant, and also wanted it to represent Chinese culture.

Japan was the first country that has broken the 200-kph barrier with its bullet train, the latest design has the shape of a platypus. France broke the 300-kph barrier in 1979, the train has a flat and low-slanting front. Chinese fastest train, the CRH3 had a world record of 350 kilometers per hour, but that was the result of technical cooperation between China and Germany. By contrast, the new CRH380 would be all Chinese product, which needs a distinctive look to match the record speed.

The design team was thinking about what has such fast speed in the world, like the horses, the mountain falls, and the rockets. They wanted to open their imagination to think about 20 unique designs. By the judgement of *aerodynamics* ability and whether they could capture the cultural soul, five designs stood out.

After the second round of comparison, one of them was the final choice. The winning model needs to look the fastest even when it is standing still. It was dynamic and has stream-line shape with *air resistance* and noise well controlled. The room for the driver and the front proportions are all excellent. After the head has been chosen, it was time to get on building it.

One of the hardest challenges was making the trains strong enough to cope with the new speeds. At 380 kilometers per hour, the air in front of the train was so compressed that it was like to keep crashing through a *brick wall*. If the head was not strong enough, the pressure alone can *disintegrate* it apart. Strengthening the thickness of shell on the train head was not the answer, the train coudn't be able to reach speeds of 380 kilometers per hour if it was too heavy. The solution was to reinforce the hollow walls to make the train very strong and very light. The welded *structs* shown in Fig. 5 – 1 can increase strength without increasing the thickness of the outer shell.

Figure 5 – 1   The welded structs in the train

380 kilometers per hour is faster than a jumbo jet taking off. The design team needed to keep the train firmly on track, and they added raised shoulders on either side of the head to control the aero-lifting force. These shoulders are like the wings, but in the opposite direction, where the air pressure forces the train down, not up. These features like the rest of train head, are gently curved, allowing the train to cut through the air with the minimum resistance. Of course, air isn't the only problem for trains to come up against. The most vulnerable part of the train is the *windscreen*, which should prepare for anything.

5.3   The Strongest Windscreen in the World

At Chinese national safety glass test center, they were looking for the world's toughest glass to fit the world's fastest trains. The test for train glass was usually under 350 kilometers per hour. It was concerned that the current high speed glass might not be able to withstand the pressure under 380 kilometers per hour. And the particular worry was birds. When the high speed train came along, there was no time for birds to escape. When the birds took off, they had hit right on the windscreen.

The first glass sample was fixed to the clamp for the test. The technicians prepared to shoot a *projectile*, which was a dead chicken. A nine-meter long cannon enabled the dead chick flying, accelerating up to 380 kilometers per hour. The result of collision was catastrophic (The glass spattered in pieces). They needed to find the glass tough enough to do the job, or the whole project would fail.

Rather than the standard glass, the next test sample, a super thin glass sandwiched with tough glass layer, wrapped around a resin core was specially designed for absorbing the high speed hit. This test was very important. If the new glass could withstand the impact, it would be directly used in the design of the train with the speed of 380 kilometers per hour. After shooting the dead chicken, the glass seemed survived. There was not a single crack under close detection. It was a perfect match for the new train. It meant train designers could go one step further to fulfill their dreams of fastest trains.

5.4 Bogie Ensures Perfect Balance of Train

At that time, the fastest rail trains had a top speed of 350kph. A jump to 380-kph may not sound much, but speed changed everything. The designers had to start many *components* from scratch.

One of the parts that most effected by high speed was the wheel unit or *bogies*. Fig. 5 – 2 shows the bogie. The painstaking process for the design of suited bogies involved endless twists and tests. It has spent eight years to perfect the design. It took this long because of a strange phenomenon called swinging.

At high speeds, if the bogie is not perfectly balanced, the whole train may shake violently

Figure 5 – 2 Bogies

from side to side, or called swinging. The swinging happens because of the surface of each wheel isn't flat. It slopes and the train could swing back and forth between these slopes, just like *pendulum*, as shown in Fig. 5 – 3. But the slopes could not be flattened, because they allow the train to turn through bends. The principles work like this: The weight of the train leans to one side as the train enters the bend, pushing the outside wheel onto the top of the slope, and the inside wheel to the bottom of the slope, as shown in Fig. 5 – 4. With the wheels running on different diameters, the train is guided into the curved path.

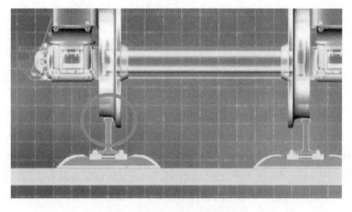

Figure 5 – 3　The slope of the wheels

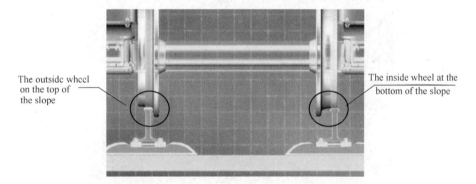

Figure 5 – 4　The situations for outside and inside of the wheels

Unfortunately on the straight line, the tiny unbalance can cause the wheels' slopes to swing. The faster the train goes, the worse swinging the train will get.

After eight years' research, the new type bogie to solve the unbalance would be conducted at the test center, under the extreme speed of 420 kilometers per hour. If there is no obvious vibration, the test is proved to be successful.

The test started. The carriage needed to stay steady on the rate to pass the test. The slightest sign of movement meant failure. The new bogie passed the test, ready for the new track and this diffculty of swinging has been overcome.

5.5 Railway Bridge to Support the Fastest Trains in the World

As tests on every aspect of new trains were continuously conducted across China, the team in charge of building the track faced a problem. They had to cross the biggest river in Asia. To solve the problem, they had to build a strong enough railway bridge.

The first Beijing-Shanghai railway was built in 1908. It has been the main route between the two cities for more than a century, but that was about to change. It would be replaced by the world's fastest railway, but to complete it, the world's busiest river has to be spanned. It proved to be the most challenging engineering of the entire project.

The vast Yangtze River divides China into north and south. At the point of the railway crosses, the river is 1.6 kilometers wide. Every day, nearly 3000 boats pass here between Shanghai and the industrial zone of *Yangtze River Delta*. Wen Wusong is in charge of this bridge.

Wen Wusong (Chief Commander of Dashengguan Bridge): This is the hardest bridge I have ever built. The deeper the water, the more difficult the task will be.

The decision that the bridge should carry six lines made Wen's job even harder, including the Beijing-Shanghai Line, the Shanghai-Wuhan Line and a subway line from nearby Nanjing, as shown in Fig. 5-5. His first job was to decide what kind of bridge to build.

Figure 5-5 The lines on the bridge

Wen Wusong: Because the water is too deep here, around 51 meters. The river traffic is so heavy. We usually use a wide span, like the highway bridge.

Wide spans as in Fig. 5-6 are ideal because it does not require a lot of underwater supports. Although perfect for road traffic, they are useless for fast and heavy trains. The most suitable design for this project would be a continuous truss bridge, but it relies on *concrete* piers to support. With water deep enough to cover a 17-storey building, the engineers had a massive challenge to construct it.

Wen Wusong: Nobody has ever attempted to build such a bridge at a depth of 51 meters.

Wen's solution was to get workers much closer to the river bed, using a submerged *platform* called *cofferdam*, which was like a giant bucket. 80 meters wide, made of solid steel and weighing 6,000 tons, each of these "buckets" was pushed to the position with no less than 5 centimeters *error* by using GPS. They provided a dry, sunken working platform, free from the *interference* of river traffic, and much closer to the bottom of the river. Engineers cut holes at the bottom of each cofferdam to accommodate steel sleeves, which extended down into soft layer beneath the river, as in Fig. 5 - 7. Then they excavated deep holes into the *bedrock* and filled them with concrete for the pillars. Finally, they filled the "buckets" themselves with the concrete so the bridge could stand on a wide platform.

Figure 5 - 6 Wide span bridge example

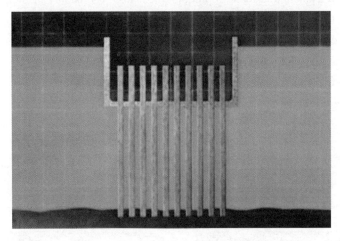

Figure 5 - 7 Cofferdams with pillars

The cofferdam has the same area as 7 basketball courts and can withstand 80,000 tons of pressure, adding six trains running at the same time. With the platform in place, the team could get on building the main structure of the bridge, using *climbing cranes* to hang the pieces in position and bolt them together. After 40 months of hard work, fixing the final bolt should be a moment of glory for the entire construction team. However, the weather at this time made the plan to change. The heat of the day had expanded the bridge so much that the bolt holes could not *align up*. They all had to wait the sunset, when the temperature dropped, leading the bridge to shrink.

Wen Wusong: We watched all holes slowly move together. When they finally matched, the whole people in site became very excited.

The completed railway bridge was an *artificial miracle*. At that time, it has the most railways, the largest span and the highest loads of any railway bridge ever in the world. However, the longest bridge record did not belong to it. This honor belongs to another section of this Beijing-Shanghai railway.

To save agricultural land, engineers had raised incredible 80 percent of the lines above the field on land bridges. It also keeps "uninvited passengers" like cows or sheep off the line. The section from Danyang to Kunshan was the longest land bridge, stretching 164 kilometers, which was four times for the *marathon* distance. This makes it to be the longest bridge at that time, twice as long as the second longest bridge in the world.

5.6 The Slabs for High-speed Track

Chief engineer Zhao Guotang was supervising an important installation——The slabs that carried the trains.

430,000 pieces of these precisely designed concrete blocks were required to complete the railway. To keep up with demand, chief engineer Zhao took the unusual step of setting up 16 manufacturing factories along the route to produce the slabs. The factory in Bengbu was the largest among them. The factory director faced a formidable challenge.

The factory director: As it is planned, the site is almost full. More than 16000 rough panels have been made and more than 13000 pieces have been polished.

The problem was that concrete alone was not strong enough to support conventional high-speed rail, and the pressure of passing trains would warp them. To battle the even greater pressure produced by CRH380, the factory adopted a two-pronged approach. The workers firstly produced a mesh steel rods as in Fig. 5-8, two of which were placed inside the mould and covered by the concrete. Lying between the steel grids was the second measure that a series of steel cables stressed by almost 500 tons forces as in Fig. 5-9. Once the concrete was solidificated, the tensioned cables wouldn't make the structure bend.

Figure 5-8 The steel grid

Figure 5-9 The pre-stressed steel

The panels were finished by a precise *grinding* machine. Lasers guided the polish stones to a tolerance of 0.1cm. The result was perhaps the smoothest, strongest and most accurate track in the world. High speed poses extreme challenges. Curves must be smoother, components must be stronger and lighter.

5.7 Speed Sensors & Central Computers

However, there is one thing impossible to upgrade——The driver. At Beijing Jiaotong University, a virtual lab measured human factors of traveling at 380 kilometers per hour. The man in charge of safety was Professor Wang Junfeng. He realized the importance for the driver to control the train. The key to safety is signaling. Professor Wang used *computer simulation* to certify why the converntional system broke down at a high speed. Under the 120 kilometers per hour, the driver is quite easy to spot the red signals. But when accelerating to 380 kilometers per hour, it is almost impossible to spot. The solution was a centralized computer that collected data about each train's speed and position via a box under the train that worked like a magcard. If a train got too close to the one in front, details of how hard the driver should brake would directly transfer to the *console*. This system only worked if it received incredible accurate information. At top speed, the *braking distance* of the new train was *enormous*, over 6 kilometers. The slightest error in the record of the speed of the train could result in a bad brake calculation, and the driver could overstep his marks by hundreds of metres. Professor Wang needed a *device* that is incredibly accurate and incredibly durable. Only one existed as in Fig. 5 – 10, but it had never been tried on a train as fast as the CRH380.

Professor Wang: This is a speed sensor (Fig. 5 – 10). There is a metal plate with 200 holes inside. Light comes from one side and there is a light sensor on the other.

The panel was fixed to the axle of the train and turned in time with the wheels. Each time a hole passed the light, the sensor would record the flash, as shown in Fig. 5 – 11. As each hole passes light, the receiver records the flash. The faster the train, the higher the flash rate will be. Producing nearly 3,000 flashes per second at a speed of 380 kilometers per hour, Professor

Figure 5 – 10 The speed sensor

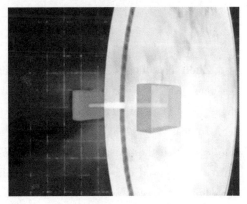

Figure 5 – 11 The principles of the speed sensor

Wang's test showed it could be perfectly recorded no matter how fast it is.

As an important part of the new system, a great leap forward in safety was achieved since steam trains.

Professor Wang said at that time, there was no relevant train control equipment, not even a *speedometer*. Now the driver can drive the train by only looking at the information of the train company. It can not only see the distance from the target point in front, but also see the current position, and it can receive instructions to know the speed that should be driven, so that it was easier to drive and can ensure the safety of the train.

5.8 The Large Roof Ensures Transferring Convenience

There was another advantage that trains can travel closer. At *peak times*, one train every five minutes filled with 1,000 people would arrive at Beijing South Station. This future main station had to be built with extraordinary specification to deal with the demands of the new railway. The Beijing South Station is enormous. It has footprint larger than Bird's Nest Stadium. More than 60,000 tons of steel were used to build it. Figure 5 – 12 shows the waiting room that could accommodate 10,000 passengers. The chief engineers of the project are Stefan Krummeck and Zhou Tiezheng.

Figure 5 – 12 The waiting room in Beijing South Station

Zhou Tiezheng: The Beijing-Shanghai high-speed railway has a top speed of 380 kilometers per hour, which is very fast. We wanted a way that people move through the station to be the same——A fast exit and a fast entry.

To deal with the half a million passengers expected to use the railway every day, they split into levels that the departure passengers are up the railway and the arrival passengers are below. This was supposed to keep the extra wide platform free even in busy times. But there was a major problem.

Stefan Krummeck: The trains of the Beijing-Shanghai line are about 400 meters long which requires long platforms. However, climate in Beijing is harsh, like cold winter, hot summer and sandstorms. So we thought that the best solution was to build a big roof, covering all the platforms to create a contained environment. But the challenge of building a big roof was that it needed supports.

Large supports took up space, but there was not any space to spare. The specialist Ho Goman helped solve this problem.

Ho Goman (Engineering Consultant Arup): For example, this side will stop the Beijing-Shanghai railway train, about 1000 passengers. A big *pillar* in the platform will affect passengers enormously. So we have to come up with a new solution.

The only option was to cram an extreme pillar between the train lines. but Ho Goman knew an ordinary design would not be strong enough. His solution was to fit this *ultra-modern* building with an ancient technology.

Ho Goman: Let's pretend that Stefan is the slim pillar to support the roof. You can imagine that once I pull him, he can fall over easily. But if Stefan spreads his legs, it is a lot harder to move him. This is same as the shape "A" frame. We don't have to add material to make it wide. It is very slim, but it meets our need perfectly. And we can put it between tracks to support the roof.

5.9 Prevent Resonance Phenomenon of Trains from Damaging

Before the train that was designed for Beijing-Shanghai railway took on tracks, the train had to pass one crucial test to avoid a mysterious phenomenon that could be a disaster for any building. This phenomenon was called *resonance*, which was a very violent form of vibration. The design team was working hard to prevent it. Fig. 5 – 13 shows the simulations models by software.

(a)   (b)

Figure 5 – 13 The simulation model for resonance research

Resonance is the tendency of the objects to vibrate most violently under the certain frequencies. For example, a pitchfork always resonates at 440Hz. The danger of resonance is the way things behave when exposed to their *resonant frequencies*.

In 1940, the newly open Tacoma Narrows Bridge in the United States suffered catastrophic damage. The vibration of the passing wind was a perfect match for the resonance of the bridge. The energy of the wind directly contributed to the *vibration* of the bridge, making it resonate so strongly that it fell apart. To avoid disasters, every part of the train was designed to resonate at different frequencies by carefully selecting materials and the way they joined together. It was essentially important for the train body and wheel bogie.

The test for the body was conducted. *Vibrating hammers* were placed at each corner on the train head. The technician slowly turned the dial, altering the frequency which they vibrated. When the frequency of the hammer was exactly matched resonance point of the train, it would vibrate very strongly. It was easy to feel the vibration but not easy to see. So a bottle of water on the train body helped to detect the vibration. The resonance point of the body had to be higher than 6Hz to keep away from the danger zone of the bogie. But it could not be too high, or it would *interfere* other components. The target was 17Hz. The experiment started from 13Hz, and the team looked for any tiny signs of resonance.

At 17Hz, the train was vibrating violently. Every surface was vibrating violently, the whole body hummed with noise, this was exact the point of resonance.

5.10 Great Engineering Significance

The new train at speed of 380 kilometers per hour was completed. Compared with the train with 350 km/h in Fig. 5-14, the new train has a longer and narrower head to reduce air noises. The body is rounder for better strength and streamline. The gaps between carriages are virtually closed for better aerodynamics. Overall, the look is slicker, faster and more aggressive.

Figure 5-14 Comparison on heads of trains at 380km/h and 350km/h

Two of Chinese most dynamic cities, Beijing and Shanghai, have been linked by the world's fastest railway. Though the CRH380's travel time was only half an hour faster than the existing 350, the less 30 minutes was worth for the effort. It has been estimated that the half an hour saved could make savings about $320 million a year.

Change was not only reflected in the economy, but also in the culture. The two big cities, one in the north and the other in the south would be connected. The culture can be blended, which made both cities stronger.

The residents in Beijing and Shanghai would benefit from the new railway, as would a quarter of the country's population living along the line. Twenty-three stations have been built, allowing families split between town and country to live a closer life. New *commuter cities* would spring up around the new stations.

The Beijing-Shanghai railway has broken records in many ways, but the changes of the lives of millions of people are its real *legacy* for the future.

Words and Expressions

railway   n. 铁路
construction   n. 建造
unprecedented   adj. 史无前例的
populous   adj. 人口稠密的
barrier   n. 障碍
track   n. 轨道
aerodynamics   n. 气体力学
disintegrate   v. 瓦解
struct   n. 结构，结构体
windscreen   n. 挡风玻璃
projectile   n. 射弹
component   n. 组成部分，组件
bogie   n. 转向架
pendulum   n. 钟摆
concrete   n. 混凝土
platform   n. 月台
cofferdam   n. 围堰
error   n. 误差
bedrock   n. 岩床
marathon   n. 马拉松
grind   v. 磨碎，打磨
console   n. 控制台

enormous   adj. 巨大的，庞大的
device   n. 设备
speedometer   n. 速度计
pillar   n. 柱子
ultra-modern   adj. 超现代化的
resonance   n. 共振，共鸣
vibration   n. 振动
interfere   v. 干涉，妨碍
legacy   n. 遗产
air resistance   空气阻力
brick wall   砖墙
Yangtze River Delta   长江三角洲
climbing crane   爬升式起重机
align up   对齐
artificial miracle   人造奇迹
computer simulation   计算机仿真
braking distance   刹车距离
peak time   高峰期
resonant frequencies   共振频率
vibrating hammer   振动锤
commuter city   通勤城市

Activities——Discussion, Speaking & Writing

1. Listening practice

Watch the video clips, write down the key words you have heard, retell the main idea in your own words and then check your writing against the Chinese scripts below.

Exercise 1：车身设计

Reference Script：加大车壳厚度并非上策，列车太重的话就无法达到 380km/h 的时速。解决对策是加固空心车壳使其非常坚固且轻盈。这些焊接的支架能增加强度，但不会增加外车壳厚度。

Exercise 2：转向架原理

Reference Script：其原理如下：列车进入弯道时重量倾向一侧，把外侧车轮推向斜度顶点，而内侧车轮推向斜度底部，由于内外车轮转动直径不同，而将列车导入曲线路径。不幸的是在直线轨道上，这些不平衡就会导致车轮斜度摆动，速度越快，摆动就越严重。

Exercise 3：铁路桥设计

Reference Script：水越深，工作难度越大。桥梁必须支撑6线铁路，这使文武松的任务更艰巨。这其中包括北京-上海线、上海-武汉线和附近城市南京的一条地铁线。他的首要任务是决定建造哪一种类型的桥梁。

Exercise 4：铁路桥桥墩设计

Reference Script：工程师在每个围堰底部挖洞以容纳钢制套筒，套筒往下伸展到江底的淤泥里，再深深嵌入岩床，灌入混凝土当作桥墩，最后将围堰本身灌满混凝土，桥梁就能坐落在宽阔的平台上。

Exercise 5：轨道板翘曲预防

Reference Script：工人先制造钢筋网，在放置两个这种钢网的模具里浇筑混凝土，夹在钢网之间的是第二道措施，一系列的钢缆以接近500吨拉力拉紧，混凝土一旦凝固，这些绷紧的钢缆会拉紧结构使它完全不会翘曲。

Exercise 6：速度传感器设计

Reference Script：传感器固定在列车轮轴上跟着车轮旋转，每个孔通过光线时接收器就记录下闪光，列车速度越快，闪光频率越高。在时速380km下每秒产生近3000次闪光，王教授的测试展示无论速度多快它都能完美记录。

Exercise 7：A形柱设计

Reference Script：他的对策是将非常古老的技术应用到这座超现代建筑上。

侯伟明：我们假设把Stefan作为一根细长的柱子支撑屋顶。可以想象我很轻易就能把它拉倒，但假如Stefan把脚张开了，要拉动他就困难得多，这就是A形架的原理。

Exercise 8：防共振设计

Reference Script：但列车在上轨道奔驰之前，必须通过一项重要测试，避免一种对任何建筑物都会造成毁灭性灾难的现象，这种现象称为共振，这是一种非常剧烈的振动形态，设计团队需要努力避免这种现象。

2. Group-work: Presentation

Group (5 to 7 members) discuss and give your opinions about the Beijing-Shanghai high-speed railway. Choose one theme below for your presentation. Clearly deliver your points of the following themes to audiences. Each group collects information, searches for literature, pictures, etc. to support your points. Prepare some PowerPoint slides.

10 minutes per group (each member should cover your part at least one or two minutes).

NEED practice (individually and together)!! **Gesture and eye contact. Smile** is always Key!!

Themes (Major topics included but not limited to)

(1) High-speed Rail Planning, Design and Construction Technology and Mode
高铁的规划设计与建造技术与模式
(2) New Technologies of High-speed Railway Operation Management and Service
高铁运营管理与服务新技术
(3) High-speed Rail Safety Guarantee Technology
高铁安全保障技术
(4) Sustainable Technology of High-speed Rail
高铁可持续技术
(5) New Technologies of High-speed Iron Energy Materials
高铁能源材料新技术
(6) New Technologies for High-speed Rail Transport Equipment
高铁载运装备新技术
(7) Railway Intelligent Transportation System Technology
铁路智能运输系统技术
(8) High-speed Railway in Integrated Transportation System
综合交通体系中的高铁
(9) Application and Practice of BIM Technology Innovation in Transportation Infrastructure
交通基础设施 BIM 技术创新应用与实践
(10) Application of Intelligent Health Monitoring System for High-speed Rail
高铁工程智能健康监测系统应用

3. Independent work: Writing

Search for the **High-speed Railway of China** in the relevant Index, like Engineering Index (EI), Scientific Citation Index (SCI) and China National Knowledge Infrastructure (CNKI). Then write a report independently on **Integrated Transportation System in China.**

Unit 6

The Channel Tunnel

Teaching Guidance

Read the article, pay attention to the **Words and Expressions** and related **sentences** and **paragraphs**.

It is a 90 miles of *tunnel*, six million *cubic* yards of *chalk*, fifteen billion dollars, and the end of ten thousand years of *separation* between England and Europe. The *Channel* Tunnel is one of the world's greatest *engineering achievements*. Perhaps the single great monument is the twentieth century *civil engineering*. This is the story of what goes on inside the Channel Tunnel.

6.1 Engineering Background

For thousands of years, the white *cliffs* of Dover have stood as England's main *defense* against the sea and against the French. Over the centuries, these chalking mountains are proved impossible. Not since the *ice age*, was there a land link between England and France, and not since Norman times, nearly one thousand years ago, has any army successfully *invaded* the island of Britain. But things changed, the cliffs are no longer England's first line of defense. The London to Dover *railway* line that once broken with *standstill* by storms at sea is now half a mile inland. But it is not the train line that moved, it stayed exactly where it was. It was England that got bigger. This is an *empire* hope, a new nature reserve *stretching out* into the English Channel. Ten years ago, this did not exist, but digging the channel tunnel generated 6 million cubic yards of chalk. They had to put it somewhere and on the English side, they put it here, at the base of Shakespeare cliff. Today, this *windswept dunes* are the main *visible legacy* of the thousands of men and women who built the channel tunnel.

The tunnel itself remains in darkness, hidden from the public eye. Richard Dance is the *maintenance* manager on the English side of the channel. He was only 22 when he *signed up* as a tunnel manager in 1987 and when digging was just beginning. He was one of the few who stayed on when the construction was complete. And he knows the tunnel better than most. Now he is driving in a service tunnel.

Richard Dance: This one is a service tunnel, which is between the two main running tunnels. We are still under ground at the moment, 7 or 8 kilometers before we start going under the sea. Actually we're retraveling from the Cheltenham *terminal*, underground to Shakespeare Cliffs where they are marine sides of tunnels. This (Shakespeare Cliff) is where they started to dig, when they dropped the shaft and boring machine to start the bore towards France. And also they dropped a small boring machine which bored towards the Cheltenham terminal at the opposite. You could drive 50 kilometers an hour with the lights off, 30 kilometers an hour with the lights on.

The channel tunnel lights on means someone is working in that section of the tunnel and traveling speeds are *restricted*. If the lights are off, the vehicles are permitted to increase their speed.

6.2 Main Tunnel and Auxiliary Tunnel

There are in fact three tunnels——two main running tunnels and one service tunnel, as shown in Figure 6 – 1. The northern tunnel carries trains running from England to France. The southern tunnel carries trains running the other way. The service tunnel is smaller and lines between the two running tunnels. It is the key to the design of the whole circuit, because it is also an *escape route* in the event of an accident, as well as it provides access for maintenance work. Cars can therefore drive from England to France, but mostly there are no *ordinary* cars.

Richard Dance: Guidance system on the vehicle actually guides the vehicle along. So you can drive it even with hands free. Two *cables* are buried in the road and the vehicle tracks two cables that control the direction. You can only operate the foot pedals to stop and go. Everything else is automated. And now we are on light off sections so we can *speed up* a little bit to 50 kph.

Figure 6 – 1 The service channel and the running channels

The service tunnel is connected to the running tunnels by a series of cross passages every 382 yards, as shown in Fig. 6 – 2. The running tunnels themselves are connected to a further series of passages called *piston* relief pipes which allow air to move between the two running tunnels. The service tunnel is not open in public. In fact, It is not open at all. The service tunnel is keeping air tight and pressure to rise. There are several reasons for this. The main one is that the service tunnel is the primary escape route for passengers in the event of an accident. Keeping it at a higher pressure than the running tunnels ensures that in the event of an accident, smoke and other pollutants will stay out. Like most builders, Richard remembers being down here is under very different conditions.

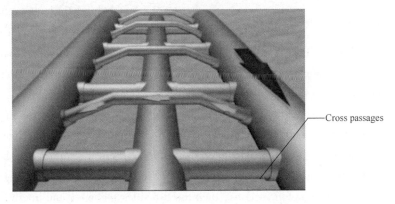

Figure 6 – 2 Cross Passages to connect running tunnels

6.3 Tunnel Digging

Richard Dance: It was very dirty, it was a totally different working environment. But you have to *guarantee* your safety of yourself. It is a very dangerous working environment with all the six tunnels being bored, three of which going to France and three going back to the land.

The *contractors* chose to start drilling from two points, one at Dover, at the base of Shakespeare Cliff, and the other at Sangatte on the French coast. From Shakespeare Cliff, six-tunnel boring machines set out, three headed to inland Folkestone and the other three headed out under the sea to France. The same happened on the French side. A few sections were built by hand.

Richard Dance: You see the channel turned from the round section to square section. This is where the boring machines were lifted out, which is the tunnel cover, Because it was dug on the surface and natural soil was put over the top of the tunnel. As we move forward, the section changes again. We go back to a round tunnel, this part of the channel was dug by hand, without using boring machine. The service tunnels now appears between the running tunnels, it *drops down* and comes along through the running tunnel and then *comes up* and finishes on the left hand of the running tunnels.

Figure 6-3 Rotating head

In France, there was no convenient cliff to cut into. Before drilling could command, they had to dig a huge access shaft at Sangatte. The boring machines were then lowered into the hole. It was hard work, the tunnel leaked, men would spend days and weeks on the icy sea water. Each tunnel boring machine was a factory. The rotating heads cut away the chalk, as shown in Fig. 6-3. *Hydraulic pistons* pushed the machine forward. Just behind the cutting head, *precast* tunnel line was *slotted* into place. The waste was taken back to shore on trains. At two points under the sea, they built huge caverns to allow trains to pass from one running tunnel to the other. The so-called cross over caverns are considered to be the greatest achievement of the tunnel builders. The English *crossover cavern* is 7 miles up to sea. The service tunnel boring machine was some distance ahead of the running tunnel machines. When it reached the side of a crossover cavern 60 yards below the ocean floor, the service tunnel machine turned down, and outward to one side. From there the tunnel was dug a series of access shafts to become crossover caverns, as shown in Figure 6-4. There they began to drill by hand: Firstly, they cleared the top sections, then the two sides sections, and finally, the two tunnels. By the time the large boring machines arrived, there was a hole in the ground with 178 yards long and 16 yards deep. With tunneling going at record speeds, the tunnels were able to take time out, to watch the two running tunnel boring machines break through into the crossover cavern on route to France.

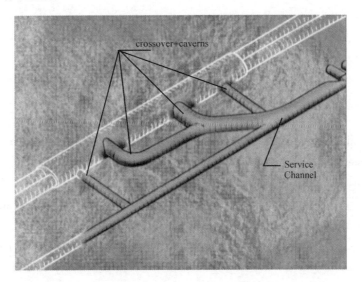

Figure 6-4 Crossover caverns

For the contractors, the crossover cavern offered the only opportunity to see the whole tunnel boring machine in action, and to keep production, the speed of machines was more than 300 yards per hour. Mostly the crossover caverns looked like two tunnels, because a *retractable* wall ran down in the middle and separated the north tunnel from the south.

When the walls were pulled back, it was possible for the train in the north tunnel to cross the south tunnel. At the French part, it could cross back again, in this way the tunnel can be divided into six sections. And any section can be *isolated* for maintenance work. To support the cavern walls, the engineers drilled *pins* deep into the chalk, they then allowed the roof to move a little for monitoring closely. When the walls were relaxed, they were covered with *waterproof* lining and *sprayed* with *concrete*.

The tunnel boring machines or TBMS meanwhile moved through and continued on their way to France. The TBMS were steered using the latest laser technology. And no stage in the 30-mile journey, was any machine leaving more than one inch of target. The service tunnel was a *breakthrough* completed in Dec., 1990, which has made history. Building became a race against time and against rising *interest charges*. Digging continued around the clock. As they drilled further out to sea, it took longer and longer for the men, tunnel line equipment and waste chalk to travel out of the tunnels.

On 28th, Jun, 1991, the final marine running tunnel breakthrough marked the end of the first stage construction, two days *ahead of the schedule* than first drafted in 1985. In the space of three short years, the contractors have drilled 100 miles of tunnel. The scale of this achievement cannot be overstated. At peak production, they drilled more than a mile per month, a rate of advance unmatched by any comparable operation before or since.

In all, 11 tunnel boring machines drilled the channel tunnel. When construction ended, they suffered various fates, a couple were sold and others were broken down for scrap, another one was sunk into deep sea and others became corporate *trophies* that were put on display outside the Euro Tunnel offices. There is also one remaining forgotten beside the highway in Folkestone. It's a *potent reminder* that digging the hole was only half the job.

6.4 Transport System

The real challenge was to turn the cave into a *profitable transport system*. It could not be any old transport system, it had to be the best. The tunnel company was already *burden by debt*. And if the debt was ever to be paid, the tunnel system has to be the first class.

It was relatively easy to link the tunnels into the British and French mainline *railway systems*. The train journey from London to Paris or Brussels now takes about three hours. However, the much harder and much steeper part was to create the *shuttle* systems for carrying cars and trucks to and from Folkestone. Not only the tunnel received the most praise of public attention, the car terminals at either end of the tunnel also represent a significant advance in transport design and technologies. The UK terminals were designed in *tandem* with the design of the carriages that would carry the cars, and the designers had many things in mind, from security to access to the *motorway*, and the most important was speed issue.

The key to the operation was that the cars, trucks and buses will be able to drive onto and off the shuttle trains. Total journey time including loading and unloading had to be kept to less than one hour, if the shuttle wants to compete with the many cross channel ferries. This meant two things——the specially designed rolling stuff and a special *platform system* at the terminals, as shown in Figure 6-5. Without these, the tunnel would not be able to compete effectively.

Figure 6-5 Special platform system

From his position at the traffic control center above Folkstone Terminal, the traffic controller is responsible for making sure cars get on and off the shuttle as fast as possible. The job sounds simple like lining up the cars and getting them on as quickly as you can. Mostly it works that 92% of channel tunnel shuttles run on time. But the system is *controversial*, passengers have to stay in the car during crossing, they have little to do and nothing to see. In the event of an accident, there's no easy way out of the *cabin*. For example, if there would a fire, passengers would firstly have to *get out of* the cars, then walk down the stairs inside the shuttle, then out into the tunnel, which they would have to walk through the dark to find a cross passage. It is especially not easy in the tunnel *filling up* with smoke.

The different trains travel at different speeds, passenger trains travel at 100 miles per hour, *freight* trains at 60 miles per hour, and passenger shuttles at 90 miles per hour. There should always be a 3 or 4 minutes gap between the front train and the back one. In any case, there should be no more than 5 trains in either running tunnel at one time. The journey time for one platform to another is 35 minutes.

6.5 Investment and Bidding

Since Roman times, Europe political leaders have wanted a tunnel linking England and France. For some, the *motivation* was commercial, for others like Napoleon, the motivation was more of a *geopolitical* nature. Following World War two, it was again trade that dominated people's thinking. In 1973, Britain finally joined the European Economic Community, the EEC, and the *conservative* government pressed ahead with the publicly funded tunnel.

2000 yards of this tunnel were dug, designed and built by many of the same people who later worked on the channel tunnel we know today, but that attempt met funding problems. A new labor government came into power in 1974 and almost immediately pulled the plug on the tunnel. The main reason was *spiraling* cost. It was to prove a *salutary* lesson. In the ten years that followed, little changed in the design of the tunnel, what did change was the method of funding. New studies suggested it was possible to build a privately financed tunnel to be paid for the future *revenues*.

In 1985, the British and French governments called for bids from companies, wishing to build and then to own the channel tunnel. And so Euro Tunnel came to win it. From the start, the company has marked itself as determinedly *Anglo-French* operation. *From time to time*, cultural differences did show themselves.

6.6 Drainage Facilities

From England to Paris, an England old train line passes slowly, which was laid for more than 150 years. It is only when it reaches the tunnel that the train really starts to pick up speed. When entering the tunnel, it's feeling like going down a hill. When they designed the tunnel, the contractors were determined to stay within the band of Chalk, also known as the chalk *core*, as shown in Fig. 6 – 6. It is the easiest way to drill through and the most stable way. *As a consequence*, the tunnel winds its way across the channel, side to side, and up and down. There are three low points in the tunnel, which act as *drainage* points for water coming in (Fig. 6 – 7). It comes as a surprise to many people that the tunnel leaks, in fact it is *deliberate*. If it did not leak, water pressure would build up *unbearably* in the rocks surrounding the tunnel, as shown in Fig. 6 – 8. The UK pumping station is 45 yards below the bottom of the sea. Above that is another 60 yards of water. Both the rock and the tunnel linings are *porous* and water *seeps* through and falls down the back of the tunnel linings. It gathers in the especially built *gully* beneath the rails and *gravity* pulls it down to the pumping stations. Every day they *pump out* more than 90 thousand *gallons* of water.

Fiure 6 – 6 The design of the tunnel to stay within the Chalk core

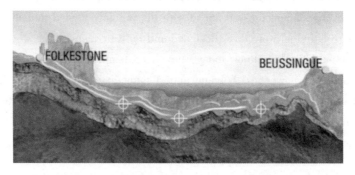

Figure 6 – 7 The drainage points for water

They don't need to pump air into the running tunnels, for that the design relies on the trains to go in there with air. Any moving object creates a flow of air behind it, especially trains. In the tunnels, this could cause air pressure to *build up*. To prevent this and to allow air to move between the two running tunnels, the contractors built a second series of connecting tunnels, as in Fig. 6 – 9. Unlike the cross passages, these piston relief *ducts* are not *sealed off*. Air moves freely between the tunnels.

The situation is different in the service tunnel, which for safety reasons is sealed off and kept at a pressure one atmosphere higher than in the running tunnels. Fig. 6 – 10 is the shaft of Sangatte in France, into which they originally lowered the tunnel boring machines. Now it has other uses, the huge fan suck in air, the air is pushed through the pipes along the *weather* ducting,

Figure 6 – 8　Water pressure surrounding the tunnel

Figure 6 – 9　The second series of connecting tunnels

down to the shaft, and into the pressurized service tunnel as shown in Figure 6 – 11. We can calculate the exact *volume* of each tunnel, the service tunnel is 30 miles long and 16 feet in *diameter*. To keep the air pressure at one atmosphere above that in the running tunnels, air must be pumped in at a rate of 220 cubic yards per second. There was a similar set of machines at the foot of Shakespeare Cliff in England.

Figure 6 – 10　The shaft in Sangatte in France

Figure 6 – 11　The air is finally pushed into the service channel

6.7　Railway Control Center

All journeys to the tunnel has been controlled and monitored from the rail control center underground at Folkestone. Michael Ben is one of the managers.

Michael Ben: The RTM is the railway control of trains and it is controlled by computers. If everything is in its right order on the correct platform and is right time for *departing*, it can be operated at any time by the RTM control on duty. And it can import *manually* dialogue commands for trains to start, go and stop or alter platforms or *alter* roads. If this person in charge did nothing in 8-hour shift, it means that all the systems work perfectly well. He or she is busy because something is going wrong or needs altering somewhere in the tunnel.

There are several types of trains going through, and each has a different number. No. 9000 represents a passenger train between London and Paris, or Brussels. No. 7000 represents a freight shuttle. No. 6000 *by contrast* represents a passenger shuttle. Right now, the passenger shuttle No. 6368 has just left Platform Seven in Folkestone. With no counter command, passenger shuttle No. 6368 has moved far away from Shakespeare cliff.

A few minutes later, No. 6368 has reached the crossover cavern. The crossover caverns are almost always closed. At night, one section of the running tunnel may be closed off for maintenance work. Passengers would hardly know what happened. Only the most seasoned traveler might notice the trains travel *marginally* slower when switching tracks. Within the service tunnel, the only evidence for train going by is a slight increase in sound above the noise of the fans.

Michael Ben: Everything of the fixed equipment is controlled from this position. The driver, for example, wanted to run into the light to be *switched on*. He would signal this to controller on this position and ask for him to light in a certain position of tunnel and there would be led by this controller who is sit here.

6.8　Fire Protection

Mostly the tunnel is kept in complete darkness. The light should only be on if there is someone working in that section of the tunnel.

A *fire brigade* passes by as maintenance manager Richard *heads for* the exit.

Richard Dance: We are just about approaching the UK pull *airlock*. It is double side doors which is designed to keep the service tunnel a slightly higher pressure than the running tunnels, which creates a safe space in the events of any incidents in the running tunnels. We'll come up and get permission to enter the airlock. The doors are open to let us enter the airlock. After the door is shut up, the doors in front will open to let us go out into the fresh air.

It takes another radio call to get the doors open.

Richard Dance: This is sixty three, we are now at the UK airlock. We ask for leaving the service tunnel.

The airlock system is slow and *tedious*, but it never fails to remind Richard how important safety was and is.

Richard Dance: If you worked with a safe manner, you could look after yourself safely. If you took chances, there are chances that things will happen.

11 men died building the tunnel, 2 on the French side and 9 on the UK side. To date, only one has died running it, but the danger is always there. And Euro Tunnel knows that everyone must be *vigilant*. With so many cars and people going through the tunnel, fire is a particular *hazard*. The fire brigade patrols the tunnel 24 hours a day. And at no time, are they under more than 20 minutes' travelling time from any point in the tunnel.

Sometimes that is not enough. One survivor on the shuttle in Nov. 1996 recalled the only accident to date happened.

The survivor: The car was absolutely full of smoke and *poisonous gases*, it is just a miracle that we were taken out from that place. If there were another six minutes, we all could walk out, but they say one or two is going to be carried out in *strtchers*.

The Nov. 1996 disaster was a serious setback for Euro Tunnel. It started on a freight shuttle a few miles out under the sea on route from France to England. Unlike the passenger shuttles, freight shuttles have open sides. When the fire started at one of the trucks, there is no way to contain either the fire or the smoke. In the ensuing confusion, many of the safety procedures failed. The driver was unable to *decouple* the trains. Fans designed to clear the smoke were not switched on, allowing smoke to be drawn back to the front *restaurant car*. 34 people were brought out and 8 were taken to the hospital, suffering from smoke *inhalation*. Fortunately, nobody was killed.

Another Survivor: In the 2 minutes, we're all on the way... on the ground to get a breath.

Fire crews and freight drivers were sent to the wrong door of the service tunnel. It took almost 400 firemen 6 hours to bring the fire under control. The heat in the tunnel was so *intense* that parts of the train were *welded* to the track.

A fireman: I've been in the tunnel for 20 years, it's certainly something on the experience ever before. My fire *helmet* although is actually intact, but when you actually have a look inside, you can see it has *blisters* (because of heat).

6.9 Maintenance Station

The section of the tunnel in which the fire happened was closed for repair for 6 months. It is estimated to have cost Euro Tunnel 420 million dollars in lost income, the repairs cost over 110 million dollars. The British government report on the fire was critical of Euro Tunnel. It is considered that the effective action by firefighters and others made the event much less serious than it might have otherwise been. A French *judicial inquiry* subsequently concluded that the fire had been started deliberately with the *culprits* not been caught. Despite the *controversy* over open sided free channels, Euro Tunnel continues to use them, even anything in new shuttles are more open sided than before. This new railroad car is about to come into service. Three years later, on the maintenance yards in northern France, the memories of the 1996 fire linger on. The Euro Star train reaches its top speed of 225 miles per hour near Lille in northern France.

To keep the ride smooth, shuttles and trains *undergo* constant maintenance and replacement programs. The sheds are long enough to *accommodate* an entire shuttle train, as shown in Fig. 6 – 12. The shuttles always move *clockwise*, and the tight turning circles at each terminal mean the wheels were out relatively quickly.

Figure 6 – 12   Maintenance shed

6.10 Tunnel Customs

As the European Union moved toward a single political and economic entity, border controls have disappeared between some member states, but they are still firmly in place between Britain and France. There is one odd effect that the border, between the two countries now occur at a range of different places. The passport control for the main line services is at Waterloo Station in London, but the British customs is at the French terminal near Calais.

The tunnel is an obvious target for *smugglers*, although the main concern for customs officials is *illegal drugs*, bootlegging of tobacco and alcohol. British taxes on goods such as cigarettes are much higher than in Europe. The customs officers use a mixture of *sound intelligence* and *gut instinct* in deciding which cars to search. But at such volume of traffic, this can never be enough. Mark Witt is a British customs officer working in France.

Mark Witt: We think that there are 95% of the smuggled vehicles traveling which are between the hours of 2:00 and 6:00. It approximately will be big time for *bootleggers* with at least 50 kilos of tobacco or the *equivalent* amount of cigarettes. So basically every vehicle you stop would have considerable quantity of *contraband*.

A dog named Catherine has been trained to look for illegal drugs. In a small car, she is looking for *heroin*, *amphetamines*, *ecstasy* and *marijuana*. The most common to find is marijuana but big of *cocaine* and heroin have also been known. A small car like this is also relatively easy to search and it doesn't take long, but a coach or a bus can take up two or three hours as the officers and dogs go *meticulously* through each section of a vehicle. Drug searches are tense tasks. The customs officer Deborah doesn't want to upset the car owner by having her dog dirty the seat. On the average at this checkpoint, they will search five or six cars an hour, twenty four hours a day, three hundred six five days a year, and still most drugs get through.

In recent months, customs officials here claim to have recovered cocaine worth more than 10 million dollars, but the tunnel hasn't made their job any easier.

Mark Witt: I'd have to say it may be harder. First of all it's another point of entry that we have to man. As opposed to the ports, we have no lists for vehicle. There is no advanced information, which makes it more difficult to target specific vehicles. And the other difficulty is that considerable of time is wasted by a staff having to travel from the UK to France in order to begin their shift. It takes approximately two and a half hours a day return just to get turn from the place of work. So that's probably very expensive for departments as well.

Customs may not like the tunnel but the customers do. The channel tunnel means that cars can cross between France and England in safety and comfort, no matter what the weather is.

Will there be a new tunnel? Euro Tunnel delivered a *feasibility* study in Dec., 1999. As before, there is much debate whether there should be a road or rail tunnel. Probably we shall have to wait some years to find out which it'll be. But one thing is for sure——We know it can be done.

Words and Expressions

tunnel   n.   隧道
cubic   adj.   立方体的，立方的
chalk   n.   白垩
separation   n.   分离，分开

channel	n. 渠道，海峡	motorway	n. 高速公路
cliff	n. 悬崖，峭壁	controversial	adj. 有争议的
defense	n. 防卫	cabin	n. 客舱
invade	vt. & vi. 侵入，侵略	freight	n. 货运，货运列车
railway	n. 铁路	motivation	n. 动机
standstill	n. 停止，停滞	geopolitical	adj. 地理政治学的
empire	n. 帝国	conservative	adj. 保守的
windswept	adj. 风刮的，被风吹扫的	spiral	v. 盘旋上升（或下降）
dune	n. 沙丘	salutary	adj. 有益的，有用的
visible	adj. 现有的	revenue	n. 收益
legacy	n. 遗赠，遗产	Anglo-French	英法的，英法合作的
maintenance	n. 维护，维修	drainage	n. 排水，排水系统
terminal	n. 终端，终点站	deliberate	adj. 故意的
restrict	vt. 限制，限定	unbearable	adj. 承受不住的
ordinary	adj. 普通的，一般的	porous	adj. 多孔渗水的
cable	n. 电缆	seep	vi. 漏，渗出
piston	n. 活塞	gully	n. 集水沟，雨水口，檐槽
guarantee	vt. 保证，担保	gravity	n. 重力，地心引力
contractor	n. 承包人	gallon	n. 加仑
precast	adj. 预制的	duct	n. 输送管，导管
slot	vt. 把…放入狭长开口中	weather	adj. 露天的
crossover	n. 转线路	volume	n. 体积
cavern	n. 大山洞，凹处	diameter	n. 直径
retractable	adj. 可取消的，可撤销的	depart	v. 离开，出发
isolated	adj. 单独的	manually	adv. 用手地，手工地
pin	n. 钉	alter	vt. 改变，更改
waterproof	adj. 不透水的，防水的	marginally	adv. 少量地，最低限度地
spray	vt. & vi. 喷，喷射	airlock	n. 气锁，气塞，气闸
concrete	n. 混凝土	tedious	adj. 沉闷的，冗长乏味的
breakthrough	n. 突破	vigilant	adj. 警惕的，警戒的
hole	n. 洞，孔，洞穴	hazard	n. 危险，冒险，冒险的事
trophy	n. （尤指狩猎或战争中获得的）纪念品	stretcher	n. 担架；v. 用担架运送
		decouple	v. 减弱
potent	adj. 有效的，强有力的	inhalation	n. 吸入
reminder	n. 提醒…的东西	intense	adj. 强烈的
profitable	adj. 有益的	weld	vt. & vi. 焊接
burden	n. 负担	helmet	n. 钢盔，头盔
shuttle	n. 短程穿梭运行的火车、汽车等	blister	n. 气泡，水泡
tandem	n. 串联		v. 使…起水泡

judicial   adj. 司法的
inquiry   n. 调查，审查
culprit   n. 罪犯
controversy   n. 争议
undergo   vt. 经历，遭受，承受
accommodate   vt. 容纳
clockwise   adv. 顺时针方向地
smuggler   n. 走私者
bootlegger   n. 酿私酒者，走私犯
equivalent   adj. 相等的，相当的，等效的
contraband   n. 走私，禁运品
heroin   n. ［药］［毒物］海洛因，吗啡
amphetamine   n. 安非他命
ecstasy   n. 迷幻药
marijuana   n. 大麻，大麻毒品
cocaine   n. ［药］可卡因
meticulously   adv. 细致地，一丝不苟地
feasibility   n. 可行性，可能性
engineering achievement   工程成就
civil engineering   土木工程
ice age   冰河时代
stretch out   延伸
sign up   跟…签订合同，签约
escape route   逃生路线

speed up   加速
drop down   突然下来，下降
come up   上来，上升
interest charge   利息
ahead of the schedule   提前
transport system   运输系统
by debt   负债
railway system   铁路系统
platform system   月台系统
get out of   摆脱，走出
fill up   （使）充满
from time to time   不时，偶尔
as a consequence   因而，结果
pump out   抽出
seal off   封闭，隔离
by contrast   相比之下
switch on   打开，开动
fire brigade   消防队
head for   走向，前往
poisonous gas   毒气，有毒气体
restaurant car   餐车
illegal drug   非法药物，违禁药物
sound intelligence   良好的情报
gut instinct   直觉

Activities——Discussion，Speaking & Writing

1. Listening practice

Watch the video clips, write down the key words you have heard, retell the main idea in your own words and then check your writing against the Chinese scripts below.

Exercise 1：辅助隧道的布置

Reference Script：每隔382码就会有横向通道，连接维护隧道和主隧道。另外主隧道还与一系列被称为活塞形管路的通道连结，让两条主隧道之间的空气流通。维护隧道不对外开放。事实上它也不是完全敞开的。

Exercise 2：隧道的挖掘

Reference Script：每一架隧道钻孔机都是一个小型工厂，旋转头切割白垩，水力活塞推动机器前进，在凿头后面，紧跟着把预制的隧道内壁放置到合适位置，废物都由火车

运回海岸。

Exercise 3：终点站设计的关键
Reference Script：运作的关键点是保证小汽车、卡车和公共汽车都能够顺利地进出穿梭列车。如果穿梭列车想要跟海峡渡轮竞争，全部的路途包括装卸时间必须控制在一个小时以内。这意味着要有两样东西——特别制造的运输工具和在终点站的月台系统的特殊设备。

Exercise 4：隧道的漏水设计
Reference Script：人们很难想象这隧道是漏水的，但其实这是故意设计的。不渗漏就无法承受隧道周围岩石中的水压。英国抽水站位于海底下方 45 码处，它的上方还有 60 码的海水。岩石和隧道内壁都是多孔渗水的。水从隧道渗出并沿着隧道内壁背部下流，汇聚于铁路下方特别建造的集水沟中，地心引力使它们流向抽水站，每天它们都要抽取 9 万加仑以上的海水。

Exercise 5：主隧道空气的流通
Reference Script：他们不给主隧道注入空气，因为设计原理是列车本身就能带入空气。任何移动的物体都会带动其后的空气流动，尤其是列车，在隧道中这可能引起气压增加。为了防止这一点，并使两条主隧道之间的空气流通，承建者建造了第二组的连接通道。

Exercise 6：维修隧道的气压控制
Reference Script：空气被推入这些导管，经由这些露天的输送管，到达通道并进入增压的维护隧道。我们能够测量每个隧道的确切体积，维护隧道一共长 30 里，直径 16 里。为了确保其气压比主隧道的高，每秒钟必须抽打 220 立方码的空气。

Exercise 7：维修站的日常维护
Reference Script：为了确保顺利行驶，穿梭列车和货运车需不时保养更换。在法国终点的维修站里，一个火车头正在接受日常维护，维修站的长度足以容下整列短程穿梭列车。

2. Group-work：Presentation

Group (5 to 7 members) discuss and give your opinions about **the features of Cross-River/Sea Tunnel.** Choose one theme below for your presentation. Clearly deliver your points of the following themes to audiences. Each group collects information, searches for literature, pictures, etc. to support your points. Prepare some PowerPoint slides.

10 minutes per group (each member should cover your part at least one or two minutes).

NEED practice (individually and together)!! **Gesture** and **eye contact.** **Smile** is always **Key**!!

Themes(Major topics included but not limited to)

(1) Selection of Bridge and Tunnel Schemes for Major Cross-River/Sea Projects
越江/跨海重大工程桥隧方案选择

(2) Study on Engineering Technology of Suspension Tunnel
悬浮隧道工程技术研究

(3) Innovation of Engineering Technology for Complex Tunnels
复杂隧道工程技术创新

(4) Industrial Construction Technology and Development Prospect of Tunnels
隧道工业化建筑技术及发展前景

(5) Construction and Latest Progress of Deep-buried Long Tunnel
深埋长大隧道工程建设与最新进展

(6) Technology of Disaster Prevention and Mitigation and Energy Conservation and Environmental Protection
工程防灾减灾及节能环保技术

(7) Intelligent Construction and Intelligent Control of Tunnel Life Cycle
隧道全生命周期的智能建造和智能管控

(8) Intelligent Tunnel Equipment and High-end Track Equipment Solution
隧道智能装备和高端轨道设备整体解决方案

(9) Application of Intelligent Health Monitoring System for Roads, Bridges and Tunnels
路桥隧工程智能健康监测系统应用

(10) Theoretical Analysis and Numerical Simulation of Deep Excavation in Geotechnical Engineering
岩土工程深开挖的理论分析及数值模拟

(11) Construction Method and Foundation Treatment in Geotechnical Engineering
岩土工程施工方法及地基处理

(12) Safety Issues, Risk Analysis, Disaster Control in Geotechnical Engineering
岩土工程安全问题、风险分析、灾害控制

(13) Application and Practice of BIM Technology Innovation in Transportation Infrastructure
交通基础设施 BIM 技术创新应用与实践

3. Independent work: Writing

Search for **Cross-River/Sea Projects** in relevant Index, like Engineering Index (EI), Scientific Citation Index (SCI) and China National Knowledge Infrastructure (CNKI). Then write a report independently on **Cross-River/Sea Tunnels in China.**

Unit 7
Chicago Underground

Teaching Guidance

Read the article, pay attention to the **Words and Expressions** and related **sentences** and **paragraphs**.

This article proposes a plan to create an underground city in Chicago. There are many examples of the underground structures to learn from. However, challenges and difficulties exist to be overcome.

Big problems like natural disasters, climate change, over crowding demand even bigger solutions. It is a *new era* of engineering, we can build our world out of any way. In Chicago, there is no room to grow, but underground city could hold thousands of peole.

7.1 Underground's Development Background

"This is the Chicago underground city express elevator, with direct connection to residence, shops and offices below the floor 50. "

Chicago of Illinois is the third most *populous* city in the United States with an iconic skyline. At the urban crowd, it streamed the daily *exclusive* city life. But Chicago's development was blocked by its own success. There are 2.8 million people here, and almost 7 million *commuters* move in and out of the city every weekday. Even the *suburbs*, once the *antidote* to the chaos of the city life, are now blocked with traffic. Despite the improvement in the *interstate* highway system, the traffic is still growing, just as all big cities. An hour will be delayed in *rush hour*.

When it comes to traffic, this "second city", only second to New York, costs the Chicago's residents nearly one week per year, and the local economy over 4 billion dollars annually in traffic.

When suburbanites reached downtown, they had to get on "L" line, one of the nation's oldest and busiest transport systems, *shuttling* at *seams*, which is unable to expand to accommodate the growing *ridership* and continuously extended rush hour. The design of the subway and L line could only accommodate eight carriages. Therefore, to increase the capacity, it would be necessary to increase the size of stations, and that would cost a great deal of money. Undeveloped land around the city is *scarce*, so outward expansion is no longer an option, and with downtown rare estate as precious, this city needs to take something *drastic*, or even *ground-breaking* solution.

Chicago already holds some of the nation's tallest buildings, this time we'll have to build downward. Imagine an underground city directly below the loop, over 1,300 feet deep, containing nearly 100 floors of apartments, offices, parks and more, which rooms enough tens of thousands of people to live, work and play. It is the best solution to the windy city space insufficiency.

It's in undertaking so big, so *unheard-of*, that all scale of the public work involved will be *incomparable*, challenging and dangerous. It will require *excavating* over 230 million *cubic yards* of rock, with construction crews laboring tirelessly below millions of Chicagoans to build a new living space with new *ventilation* and lighting. To ensure the underground city of Chicago's success, it won't be easy, but it may be necessary. Expanding underground is a solution that has been becoming more acceptable, with emergence of new techniques to make living underground a reality.

Chicago was an ideal place for this bold experiment. From the ashes of the great fire in 1871, the city grew to become the center of the *architectural* revolution, establishing the first *skyscraper* in 1885, and hosting the most influential World's Fair in history in 1893. By building underground, Chicago had its reputation as one of the world's most *forward-thinking* cities on the planet. But it may not be the first.

7.2 An Example for Underground under Amsterdam's Canal System

Amsterdam has already been proposing space below ground, since there is no land above. In this Dutch city, *streetcars*, *pedestrians* and 550,000 bicycles, compete space in a narrow street, and the volume of cars is a growing problem.

Bas Obladen (SR. consultant, Struklon Engineering): Because Amsterdam is a nice historical town, and there is a lot of traffic in the town, you can not enjoy the town because there is everywhere of traffic. There is traffic on the channels, there is traffic on the *quays*, everywhere there is traffic.

Ancient city planners never foresaw these beautiful *winding pathways* would be overrun with *automobiles*. Today there is a movement to get cars off the streets. But to do that, while still maintaining the Amsterdam's historical *charm*, it has to rely on its signature feature of its historical past——The city's *canals*.

The plan, a network of underground complexes, built beneath the city 65 miles of iconic waterways. Parking *garages*, movie theaters and shops, were all built beneath miles underground, above which it just happens to be full of water.

Bas Obladen: Working in the canal space is really indeed the best options, because you make from one square meter which has no value, into 6 square meters which has a lot of value, of course.

In fact, the plan will convert over 20 square miles of largely *unutilized* space, and provides *storage* for thousands of cars. But building in Amsterdam requires sensitivity to city's unique *architecture*. Amsterdam's streets can't contain construction equipment, some are so narrow that massive trucks needed to build underground won't even fit.

Even if crews could access the area from the sidewalks, there would be dangerously close to the city's narrow canal houses, *precariously* stood above the waterways.

"Let's say in Chicago, you have to *dig* a hole, and you can place a house, it's a complete other way of building. In Amsterdam, it's not so easy to build a house, because of the soil is what someone calls soft water."

Amsterdam's soft water or swamp soil makes constructing large structures nearly impossible, which wide buildings are rarely higher than three or four *storeys*. Most of the canal houses are supported by little more than *wooden piles* standing into that soft soil, as shown in Fig. 7 - 1. Heavy *vibration* generated by construction could spell disasters. Only the canals are far away from the buildings and wide enough to fit construction equipment. But the canals are full of water, so engineers have proposed the solution as unique as the city itself.

Figure 7 - 1　The wooden piles supporting the canal houses

Amsterdam's canal system has 16 working locks, which can be closed at any time. With the canal closed, the canals can be drained, as shown in Fig. 7 - 2.

Figure 7 - 2　Construction after the lock was closed

"I can imagine that people think that most difficult thing is to bring out the water, out of the canals. But in fact, that is technically one of the easiest things."

Once the canals are dry, crews will get to work, digging 30 feet down before sealing off a new surface that both serves as the roof of the underground complex and the floor of the above canal. It is a solution that not only minimizes the risk to the thousands of buildings beside the canal, but also keeps the city from having to shut down.

Bas Obladen: Because we are working in the canal space, there is less inconvenience for the people.

Later, restoring the canals is simply conducted by opening the lock gates.

Bas Obladen: After we have finished it, the canal is exactly the same as it is now, only there are no cars on the ground anymore, so you can enjoy the city and the nice historical houses.

The project would move thousands of cars from sidewalks. And if it works, it could spur other cities to take up the challenge. There is just one problem that few cities have canals from which to access their downtowns. In Chicago, there's no way to dig up the streets without *paralyzing* the city and *infuriating* millions of residents. To get a better Chicago, engineers have to think deeper to find out ways. They have to look for solutions beyond the city's borders.

7.3 Excavation of Various Layers of Geology

Chicago may be the perfect location for this *subterranean* experiment, because deep beneath the glass, steel and *asphalt* of this historical city, lies the even richer *geological* past.

Charles Dowding (Professor of Northwestern University): I think it's easy to imagine Chicago's geology being a *layer* cake, made from various different geologic materials.

The first layer is *approximate* 100 feet of *glacial clay*, which is loose-soft soil that is not suitable to afford large structures. But beneath that lies 400 feet of *limestone*, which is strong rock that's perfect for excavating large caves.

Charles Dowding: The limestone beneath Chicago is an advantage, because it is stronger than *typical concrete*.

That layer below the limestone is 100 feet of *shale*, which is unstable rock that is less ideal for building. Shale is five to ten times weaker than limestone, it is subject to this phenomenon we call *slack*. When it is exposed to water, it will expand and *fracture* in debris. *Vertical* access *shafts* could be drilled to the shale, to the next layer of limestone, which extends 400 feet. It is a pattern that repeats again down to 1300 feet, before hitting a deep layer of *sandstone*, as shown in Fig. 7-3.

Figure 7-3   Repetitive pattern until to the sandstone

Charles Dowding: The top of sandstone is very weak, so weak that it can be *penetrated* by *fire hoses*.

That weakness means that the underground city could go no deeper, but it is the alternating sections of limestone that make Chicago an ideal location for this hard geological engineering. Excavating limestone is something that Chicago was already familiar with. At Volkan Mine in Naperville, Illinois, 45 minutes from downtown Chicago, a limestone and *dolomite mine* has been in operation for more than two years, with enough material available for decades more.

7.4　Effective Technology for Excavating Stratum

Crews were sent 400 feet deep to excavate the latest level of the *rock stratum*, their goal was to move thousands of tons of rock in one day. They did the process by the approach called "*drill and blast*". A 385 *horsepower* rig can dig 16-foot long holes in the rock and 59 holes per time, which were filled with *explosives*, carried out by an *anfo loader*, as shown in Fig. 7-4. Two tanks on this truck held five thousand pounds of anfo, which are *ammonium nitrate* covering the *fuel oil*. This explosive material was injected into the holes via an *air hose*, then set the *detonate* sequence.

Charles Dowding: It's a slow *methodical* process. Drilling and blasting must be carried out in strict sequence, which involves drilling, loading, evacuation, blasting and ventilation. Difficulty in transporting personnel during evacuation limits the completion to a minimum four to eight hours per round.

Safety is the primary concern, but this method got results. The 1100 pounds of anfo used in this round, dislodged 1,400 tons of

Figure 7-4　The rig for "drill and blast"

material, and excavated a 14-foot layer from the thick wall. Drilling and blasting created an open space, where once there was a solid massive rock. It also created structural engineering challenge which must be addressed immediately.

Charles Dowding: Excavating large openings can cause the stress concentrations that can overload the rock without *reinforcement*, and the larger and deeper the cave is, the larger the stress concentration will be.

To strengthen up, another drill on the job bored vertical holes, then filled with *epoxy* and steel bars to reinforce the *ceiling*. After the explosion, excavated materials would be moved out by a conveyor belt and trucked to nearby construction sites to be used as "*aggregate*", which is a crucial *ingredient* in concrete, that was put to immediate use in Chicago. This material is limestone that can be used to fabricate concrete, then it can be reused to the underground city again. The underground city could use some of material, but not all of it, leaving enough extra material to pave over 30,000 miles of roads, which could stretch around the globe one and a half times. With the easy excavation, digging out a new city is not a problem, but holding the entire existing city is a challenge.

7.5　Access to the Underground without Disturbing the City Life Above

Chicago's tallest building at that time, the Sears Tower, *soars* to 108 storeys of 1,450 feet into the air, held up by 200 piles extending nearly 100 feet underground, sending the building's load into the surrounding rock. And it is just one building. Beneath Chicago lies a forest of *foundations* and *piles*. If digging too close to them, the buildings they support could be in trouble.

"If you want to dig a hole directly under the end of the pile, it could cause the building collapse."

For safety sake, the first level of the underground city must be positioned hundreds of feet beneath the lowest pile, but getting there without closing the city streets, rerouting miles of train lines, or redirecting *drainage* channels, water pipes and *utilities*, will be an incredible challenge. Avoiding them all together will be the easiest and cheapest option.

Charles Dowding: To access the underground city, it is important to construct in area where there are no constructions and utilities.

Unable to disturb the city life above, crews would have to begin their access to the work site from miles away. Starting from outside the city center, crews and equipment would tunnel downtown, ultimately reaching about 300 feet below the center in Chicago, which is deep enough to avoid building foundations and urban *infrastructure*. There they would undertake the drilling and blasting for the construction of the first level, and *stage* area for building even deeper. But as they do, they have to ensure the walls of these new underground cities are strong enough to hold back the water of *mightiness* of Lake Michigan.

7.6 Underground Construction in Moscow

Chicago is famous for two things ——it's big and cold. However, there is one place that bigger and colder —— It is Moscow. With 15 million people in the *metropolitan* area and over 3 million cars causing some of the world's worst traffic, Moscow's explosive growth is putting it on a city to the *brink* of disaster.

Mikhail Mikhailovich (Architect): The real first priority right now is to get rid of the transportation collapse system in the city and the constant traffic jams.

To cope with the traffic problem, the capital of Russia had no way to go but to go down.

Mikhail Mikhailovich: Moscow is a historic center, if we want to build anything without ruining any architecture, we have to build underground.

At that time, Russia was coming in an underground construction boom, with two new projects that aimed to prevent the city from being on crisis. The first, Tovoskaya Mall would be the biggest underground mall in Europe when finished with 4 storeys deep, spreading over 350,000 square feet. But building it meant redirecting nearly 294,000 cars that quick through adjacent streets each day. And crews must work perilously close to traffic. The second site, Paul Leitz Square would one day be a five-storey deep shopping and parking complex, *integrating with* the train station above it. But before this plan was realized, crews must cope with the city's soft, muddy soil.

Mikhail Mikhailovich: Moscow has a very different and difficult soil combination, therefore, every construction requires a very serious range of studies.

The soft soil doesn't pose problems when erecting a shorter building that represents the city's signature, but it's not ideal for building underground. Just to get a firm enough soil for building means digging down over 90 feet into *stiffer* and stronger *clay*.

Mikhail Mikhailovich: One thing I am certain is that it is not very difficult to build underground city in Moscow, we have all kinds of necessary equipment and enough knowledge to do it, because we started building at the beginning of 20th century.

One of the *treasures* of this historic city is the Moscow Metro in hundreds of feet below ground. The city's underground rail system is one of the busiest in the world. Every day 9 million people ride 4,000 trains and pass through 177 stations over 190 miles of track. The metro was constantly expanding to keep up with demand. The newest station is Park Pobedy, opened in 2003 after 16 years of construction, which possesses many amazing achievements. The longest escalator takes four minutes to transport passengers to a platform 328 feet below ground, which is the deepest railway station at that time.

To the construction of Paul Leitz Square, the crews knew these challenges all well. Digging one giant hole for underground mall was not *impracticable*. Because Moscow's loose and soft soil was so hard to hold back. So instead of digging space for the complex and then building concrete walls around it, they were doing the opposite——Building the walls at first was conducted by an excavator in Fig. 7-5. Six sets of long *claws* dug into the swamp, cutting a two-foot wide *chamber* that was immediately filled with concrete. Then forming a wall around was conducted, followed by *hollowing out*. It was slow and *laborious*, requiring frequent stare and stop just to clean.

Figure 7-5 The excavator to build the wall

Once the wall was set, crews could continue digging, but challenge didn't end after walls were in place. Digging in the city meant rerouting utilities, *drainage systems* and other infrastructure. Eventually, crews would be able to hollow out the section of the Paul Leitz Mall. Like on the other side of the construction site, all crews were under 100 feet deep, strengthening the walls with steel rebar, concrete and *plastic*. With so much water *dripping* through, the danger of collapsing inward was all too real. That was why these giant poles as in Fig. 7-6 with five feet in diameter and about 200 feet in length were put into hole to hold back walls. Despite their massive size, they were temporary, which would be removed once the entire box was reinforced. Because water always finds the weakness to pass, engineers must be *vigilant*, even though Paul Leitz Square is half a mile from Moscow River.

Figure 7 – 6   The giant poles to hold back walls

Back to Chicago, the city *perched* on the shores of the Great Lake, how to stop this lake water from getting into the underground construction would be a big problem. Allowing water to sink in would be a disaster. Each level of the Chicago Underground City would have to be painstakingly reinforced. In order to stay ahead of population boom that threatens the entire city, the underground city would have to be built quickly.

7.7 Ventilation Technology for Underground Engineering

Before the underground city in Chicago was completed, there was a huge issue to be solved. In any square, air quality is the priority. During construction, engineers can employ ventilation methods in use of the nearby Volcan Mine. The vertical shaft works to pull air in from mine's access tunnels and directly up, making sure dust and other dirt don't *linger* for too long. This is a system meant to provide comfort and security for *dozens of* miners working ten hours' shift. Replacing miners with thousands of residents living underground 24 hours a day, the ventilation becomes far more complex.

Charles Dowding: People will have to occupy in this space for 50 to 100 years, while miners are in the mine for less than ten years.

Deep beneath the ground, there will be no opened windows, no *breezes* to keep fresh air circulate. However, the ventilation problem has already been solved, far beyond the needs for any underground city.

In Boston, Massachusetts, engineers behind the "big dig", which was the most expensive public work project in US history were faced to thousands of cars passing through almost four miles of tunnels every day, exhausting tons of poisonous gases. Moving cars can push the air through the tunnels, but traffic jams can cause air to linger and dangerous gases to build up. So the Massachusetts *turnpike authority* installed a complex ventilation system to keep the air moving. Near tunnel openings, *jet fans* circulate air. But with some tunnels as deep as 120 feet, Boston had to take more *extreme* measures to bring air underground. Deep beneath the building in heart of Boston, lies the *lungs* of the centural *artery* tunnel system——A building of 125 feet tall pulls in fresh air and transfers it to tunnel's bottom.

Helmut Ernst (Chief Engineer, MTA): The intake air will be sucked in through the *skylights*, and sucked down to the chamber, through a series of pipes into the tunnel.

One hundred feet below ground in a massive intake room, giant fans pull fresh air down the shaft. Four hundred horsepower engines are capable of moving 333,000 cubic feet per minute and generating wind's speed up to 62 miles per hour. It's a powerful system, based on age-old techniques.

Chicago Underground City will apply much technology used in Boston at ten times depth. Underground city's walls and floors will be filled with ventilation chambers. Fans at the bottom of the city (Fig. 7 – 7) will pull fresh air in through *intake* shafts located at the top, where the air is sent down 1,300 feet, and circulate throughout the city. Then the air is returned to the surface to be exhausted. Regulating air quality for 24 hours for residential population is

Figure 7 – 7 Fans at the bottom of the city

not just automobile traffic, it requires more fans and a control system that would be the eyes and ears of the underground city. It's something that provides security and peace of mind for the residents below Chicago.

7.8 Advantages and Disadvantages of Living Underground

With the concept of underground city has been formed, one factor of Chicago's climate will be regulated.

Charles Dowding: It is advantageous to live underground. For instance, you are not subject to vigorous winter of Chicago.

Deep beneath the city streets, residents wouldn't have to worry about snow or sun, because there is one thing about living underground predictable——It is the temperature. No matter how hot or cold the ground is in Chicago, 22 feet underground is around 55 degrees Fahrenheit at any time of the year. But the deeper you get, the warmer it becomes, thanks to something called "*geothermal gradient*". On the surface, temperatures are affected by the sun and atmospheric changes, but at 200 feet underground, the temperature remains constant, then for every 100 feet deeper, the temperature rises by about one degree Fahrenheit. Fortunately, in the underground city, the geological limits of construction would ensure that things would not get too "*roasted*".

Charles Dowding: I think it's quite nice to maintain a constant temperature of 65 to 68 degrees Fahrenheit, much better than the extreme temperature difference between zero and a hundred degrees Fahrenheit on surface.

A constant climate is one advantage of underground city compared to the surface level, but there is one quality which the underground city can't hold.

Charles Dowding: I think the most important issue for the underground city would be the lighting itself. Because without artificial *lumination* below the ground, it is completely black.

Daylight is so vital to human health and happiness, and only comes from one source. Even in Moscow, with just 70 sunny days per year, the underground mall Okhotny Ryad does all what it can. With skylight that decorates roof of mall, allowing sunlight to travel three storeys underground, it is a solution that can work in Chicago too, *apart to a point*. Massive skylight will *illuminate* construction's *upmost* levels, but deeper down, there will be no way for sunlight to reach people. This simple fact will make people choose to live on the ground, inspite of the Chicago's cold, worth to many people.

7.9 Simulating the Day and Night Lighting System

Answers to the lighting might be found at Rensselaer Polytechnic Institute's Lighting Research Center in Troy, New York. Here, researchers focus on light effects on human health and *circadian system*, which is the 24 hours' cycle that all humans are programmed to follow.

Mark Rea (Ph. D, Director, Lighting Research Center): The circadian system wants to see a regular 24-hour pattern, it's a pattern that we've *evolutionarily* grown up with. It has been 100 years since we had electric light, allowing to extend the light into the night time, and also it happened since the last 100 years that we lived indoors.

With moving indoors, offices and homes are equipped with electric lighting, humans have disrupted the light and dark pattern that regulates the circadian system.

Mark Rea: We can be in perfect good light to see, but we can be biologically dark. You can't necessarily stimulate circadian system with electric lighting.

With only artificial light, the internal clocks of underground residents would go disordered. The circadian system clock runs off time any way. If you don't have reset the circadian clock, *synchronizing* with sunrise and sunset, you can go into "construction *jet lag*" as we call.

To maintain the synchronization, engineers will have to design a lighting system that *mimic* patterns on the surface, but even this artificial means doesn't work far enough.

Mariana Figuerio (Ph. D, program director, LRC): It is important not just how much light you get, but when you get the light and what kind of light you get in.

In the lighting research center lab, a test measures lights effect on brain activity. *Hooked up to an EEG* (electroencephalo-graph), a *testee* is exposed to different *wavelengths* of light. While all wavelengths stimulate the brain, blue light is the most effective.

Mariana Figuerio: The reason why we are working with blue light is because we have shown that the circadian system is *maximally* sensitive to blue light.

This fact may be because the color of the sky has guided our brains to respond to blue visible light. Rensselaer Polytechnic Institute has found the need for blue light is so *in grain* in human DNA, that it only takes a small amount to trigger alertness.

Mariana Figuerio: So this way you can effectively achieve the same result in reducing *melatonin* level, or increasing brain activities, with less amount of power or less light. So it is an *energy-efficient* way of impacting circadian system.

That energy-saving way could dispense the city from having bathed common areas in blue light, or by an ultra hot light to simulate sunlight. Instead, residents could take a more personalized approach.

Mariana Figuerio: Blue light is so effective for circadian system. A very low level of blue light could be able to affect circadian system.

Brief exposure to blue light is as effective as using the common way for waking up. It's just like a cup of coffee to be sober, you can use blue light for 15 minutes or 20 minutes to wake up, to give you some alerting *stimulus*. This field of study is so new that the researchers are just beginning to apply these findings to modern architecture.

Mark Rea: When we design lighting for buildings, we don't design deliberately to use a lot of blue light. So even though in the office, we can see well. We could be in what we call a circadian darkness.

Rensselaer Polytechnic Institute's breakthrough in light research will help city planners overcome some huge psychological challenges of living underground, by making the environment not only pleasant to look, but also pleasant to live in.

Charles Dowding: There would be lots of opportunities to simulate the light. In such ways, you can provide scenes of daytime and nighttime, you can provide scenes of changes to make life enjoyable and attractive.

If we understand the system, we can design electric light to do everything, as natural light does. There is no technical barrier to provide humans with regular day and night pattern. Of course, nothing can compare to nature, but it is something engineers could design out. With all the comfort same with the surface world, Chicago's Underground City could be home to tens of thousands of people.

7.10 Concerns to Live Underground

With more construction challenges overcome, it is not only life below Chicago that would begin to improve, but also the life on the ground changes. At street level, the city is no more crowded with less traffic, and transportation system is more able to go. People are no longer commuting from the outside of the city, but from bottom to up.

Charles Dowding: The city can be more beautiful if most of the facilities are placed underground. Placing major facilities underground allows us to make more proper use of the surface land.

In terms of Boston, after the big dig, the city has been *reunited* again, some surface land has been *reclaimed* for public use, rather than building *elevated highway* above it. There are parks where there used to be steel structures. These ground improvements could go a long way to support to expand the underground city, but continuing to dig beneath Chicago's skyscrapers could be met with oppositions, so the underground city will have to navigate uncharted territory.

Charles Dowding: The state of Illinois owns the whole lake, from the side outward, and there is no building. Perhaps starting by the lake shore will be much less expensive and more quickly to realize the activation of the underground city.

By connecting a new network of underground chambers to the existing underground city, urban planners could create a mirror image of the city beneath the lake.

Charles Dowding: To change the way the world thinks about Chicago, it may move the current beautiful vertical skyline and think about the underground skyline.

There's just one problem. The first part of the underground city would be built without anyone living there. If the urban planners wanted to expand, the challenges that once only defeated the crews, could now impact tens of thousands of permanent underground residents. It is impossible to see the complete picture of underground like the surface land. So there are more unknowns to hit during construction.

The most important issue to address is the natural gas that has been located underground for a long time. If crews accidentally *encounter* the hidden gas line, the consequences would be *disastrous*.

Evacuation plans would share much in common with the existing skyscrapers.

Charles Dowding: The situations for tall buildings like how many elevators will be necessary, how many stairways for escape will be necessary are same discussions that would be applied to underground facilities as well.

But thousands of feet below ground, these considerations would be turned upside down. With elevators unworkable, the stairs will be the only means of escape. Those people at the bottom level would have to climb the equivalent of 100 storeys buildings to escape, which is not an option for the very young, elderly or *handicapped*. Emergency would require fast thinking and fast acting on the *municipal* authorities by *blockading* trouble's areas, as in Fig. 7-8, protecting the rest of the city by discharging the dangerous gases.

Figure 7-8 The fast action to blockade the trouble areas

The technical challenges of building underground may be overcome. But the psychological challenge is another factor. The question is how much time people would spend on accepting to live underground. But with fact of the city's bursting and struggling the keep-up population boom, those concerns may be *overridden* by practical necessity.

One day people will no longer just shop, or drive underground, one day we just made to live there.

Words and Expressions

populous adj. 人口稠密的
exclusive adj. 独有的;n. 专属
commuter n. 通勤者
suburb n. 郊外
antidote n. 解毒剂,解药
interstate adj. 洲际的
shuttle v. 往复运动,梭动
seam n. 缝合处,夹缝
ridership n. 乘客数,客运量
scarce adj. 缺乏的,稀有的
drastic adj. 激烈的,猛烈的
ground-breaking adj. 独创的,开创性的
unheard-of adj. 空前的,前所未闻的
incomparable adj. 无可匹敌的
excavate v. 挖掘
ventilation n. 通风设备,空气流通
architectural adj. 建筑学的,建筑上的
skyscraper n. 摩天楼
forward-looking adj. 有远见的
streetcar n. 地面电车
pedestrian n. 行人,路人
quay n. 码头
automobile n. 机动车
charm n. 魅力,魔力
canal n. 运河
garage n. 车库,修车厂
unutilized adj. 未使用的
storage n. 存储,仓库
architecture n. 建筑学,建筑风格
precariously adv. 不安全地,不牢靠地

dig v. 挖掘
storey n. 楼层
vibration n. 振动,震动
paralyze v. 使麻痹,使瘫痪
infuriate v. 激怒,使大怒
subterranean adj. 地下的,隐蔽的
asphalt n. 沥青,柏油
geological adj. 地质的,地质学的
layer n. 层,层次
approximate adj. 近似的,大概的
limestone n. 石灰岩
typical adj. 典型的
concrete n. 混凝土
shale n. 页岩,泥板岩
slack adj. 松弛的,疏忽的
fracture v. 破裂,断裂
vertical adj. 垂直的,立的
shaft n. 竖井
sandstone n. 砂岩,沙岩
penetrate v. 渗透,穿透
dolomite n. 白云岩
mine v. 开采,采掘
horsepower n. 马力
explosive n. 炸药
detonate v. 引爆,使爆炸
methodical adj. 有系统的,有方法的
reinforcement n. 加固,强化
epoxy n. 环氧树脂
ceiling n. 天花板,上限
aggregate n. 骨料

ingredient   n. 原料,要素
soar   v. 高耸
foundation   n. 基础,房基
pile   n. 桩
drainage   n. 排水,排水系统
utility   n. 公用事业,公共设备
infrastructure   n. 基础设施,公共建设
stage   v. 搭建
mightiness   n. 强烈,强大
metropolitan   adj. 大都市的;n. 大都会
brink   n. 边缘
stiff   adj. 坚硬的,僵硬的
clay   n. 黏土,泥土
treasure   n. 宝物,瑰宝
impracticable   adj. 行不通的,不能实行
claw   n. 爪子;v. 抓
chamber   n. 空间
laborious   adj. 勤劳的,艰苦的
plastic   adj. 塑料的;n. 塑料制品
drip   v. 滴下;n. 水滴
vigilant   adj. 警惕的,警醒的
perch   v. 栖息,位于
linger   v. 徘徊,缓慢度过
breeze   n. 微风
extreme   adj. 极度的
lung   n. 肺
artery   n. 干道,主流
skylight   n. 天窗
intake   n. 通风口,引入口
geothermal   adj. 地热的,地温的
gradient   n. 梯度,坡度
roast   v. 烤,烘
lumination   n. 照明,光照量
illuminate   v. 照亮
upmost   adj. 最高的,最重要的
evolutionarily   adv. 进化上地
synchronize   v. 同步,同时发生
mimic   v. 模仿;adj. 模仿的
EEG(electroencephalo-graph)   abbr. 脑电图描记器
testee   n. 应考人,测试对象
wavelength   n. 波长
maximally   adv. 最大地,最高地
melatonin   n. 褪黑激素
energy-efficient   adj. 节能的,高能效的
stimulus   n. 刺激,刺激物
reunite   v. 使再结合
encounter   v. 遭遇,偶然碰见
disastrous   adj. 灾难性的,损失惨重的
handicapped   adj. 残废的;n. 残疾人
municipal   adj. 市政的,地方自治的
blockade   v. 封锁;n. 阻塞
override   v. 推翻
new era   新纪元,新时代
rush hour   交通拥挤时间,高峰时间
cubic yard   立方码
winding pathway   蜿蜒小径
wooden pile   木桩
glacial clay   冰川黏土
fire hose   消防软管
rock stratum   岩层
drill and blast   钻炸法
anfo loader   铵油炸药装填器
ammonium nitrate   硝酸铵
fuel oil   燃料油,燃油
air hose   空气软管,通风软管
integrate with   使与…结合
hollow out   挖空
drainage system   排水系统
dozens of   几十,许多
turnpike authority   收费公路管理局
jet fan   喷射鼓风机
apart to a point   在一定程度上
thermostatic system   恒温系统
circadian system   生理系统
jet lag   时差,时差感
hook up to   连接
in grain   彻底

elevated highway 高架公路

Activities——Discussion, Speaking & Writing

1. Listening Practice

Watch the video clips, write down the key words you have heard, retell the main idea in your own words and then check your writing against the Chinese scripts below.

Exercise 1：运河系统优势

Reference Script：阿姆斯特丹的运河系统有16道闸门，可以随时关闭。闸门一关，运河里的水就能排干。等运河一干，施工人员就可以开工了，往下挖30尺后封闭新的表层，将同时作为地底建筑的屋顶和运河底，稍后只要打开闸门就能让运河恢复。

Exercise 2：钻炸法流程

Reference Script：这个385匹马力的钻头能在岩石中掘出16尺孔洞，每次掘59个洞，洞里将填入火药，由铵油炸药装药机负责执行。这辆卡车上的两只槽中装有五千磅的铵油炸药，也就是包覆着燃料油的硝酸铵，这些炸药经由空气软管注入洞内，接着设定循序引爆。

Exercise 3：建造混凝土墙

Reference Script：所以他们不是先凿洞，然后在四周搭建混凝土墙，相反的是，他们先把墙砌起来。工作由这台开挖机负责展开，六对长爪挖进泥泞中，在挖出二尺宽的洞后，马上灌入混凝土，四周形成墙面后，接着再挖空中间部分，这个过程缓慢又费工，经常边挖边停只是为了清理泥泞。

Exercise 4：地底通风技术

Reference Script：地底城的墙壁和地板将布满通风机房，城市底部的风扇把新鲜空气从顶端的入风井抽入，空气将送入1300尺地底，在整座城里循环，之后送回到地面排出，持续24小时为所有居民调控空气品质。

Exercise 5：恒温系统的优势

Reference Script：全拜"地热梯度"所赐，地面的温度会受到太阳和大气转变的影响，但是在200尺的地底，呈恒温状态。每向下100尺，温度会上升华氏一度左右，幸好在地底城里，由于地质对工程建设所造成的影响，可以确保一切不被烤焦。

Exercise 6：紧急状态的措施

Reference Script：当电梯无法运作时，楼梯将成为唯一的逃生路径。在底层的人们必须爬到等同100层的楼梯逃生，对年幼、年长或残疾人是不可行的。而紧急状况依赖于市政府当局快速的决断和反应，包括封锁灾害区域，保护城市其他区域免受有毒气体的危害。

2. Group-work: Presentation

Group (5 to 7 members) discuss and give your opinions about the features of Underground Engineering. Choose one theme below for your presentation. Clearly deliver your points of the following themes to audiences. Each group collects information, searches for literature, pictures, etc. to support your points. Prepare some PowerPoint slides.

10 minutes per group (each member should cover your part at least one or two minutes).

NEED practice (individually and together)!! **Gesture** and **eye contact**. **Smile** is always Key!!

Themes (Major topics included but not limited to)

(1) Development Planning of Underground Space Construction
地下空间建设发展规划

(2) Construction Technology of Underground Building Engineering
地下建筑工程施工技术

(3) Waterproofing Technology of Underground Building Engineering
地下建筑工程防水技术

(4) Fire Prevention Technology of Underground Building Engineering
地下建筑工程防火技术

(5) Lighting and Ventilation in Underground Space
地下空间照明与通风

(6) Underground Engineering and Tunnel Security Construction
地下工程与隧道的安防建设

(7) Risk Assessment and Management in Underground Engineering
地下工程中的风险评估和管理

(8) Intelligent Health Monitoring System for Underground Engineering
地下工程智能健康监测系统

(9) Theoretical Analysis and Numerical Simulation of Deep Excavation in Geotechnical Engineering
岩土工程深开挖的理论分析及数值模拟

(10) Construction Method and Foundation Treatment in Geotechnical Engineering
岩土工程施工方法及地基处理

(11) Safety Issues, Risk Analysis, Disaster Control and Management in Geotechnical Engineering
岩土工程安全问题、风险分析、灾害控制及管理

(12) Physics and Numerical Simulation in Geotechnical Engineering
岩土工程物理及数值模拟

(13) Design Calculation and Prediction Method of Geotechnical Engineering

岩土工程设计计算及预测方法

3. Independent work: Writing

Search for the famous **Underground Engineering of the World** in relevant index, like Engineering Index (EI), Scientific Citation Index (SCI) and China National Knowledge Infrastructure (CNKI). Then write a report independently on **the Latest Progress of Underground Engineering in China.**

Unit 8
Hong Kong's Ocean Airport

Teaching Guidance

Read the article, pay attention to the **Words and Expressions** and related **sentences and paragraphs**.

Where do you put a huge new airport in a jam pack city like Hong Kong? The solution called for one of the biggest construction projects of the twentieth century.

Just how crazy would you have to be to build one of the world's busiest airports in the middle of the sea and the heart of the *typhoons*?

It's a crazy idea that only became this due to some clever engineering techniques, including a world war two *bomber*, a cold war spy bug, an eight-hundred-year-old water *pump*, a *brass band* and a *vintage* racing car.

8.1 The Busiest Airport in the World

Most people who pass through Hong Kong *Chek Lap Kok Airport* have absolutely no idea what it took to build it. The land of Hong Kong new airport had to be created. It was one of the biggest *reclamation* projects of the twentieth century. And it couldn't have been done without a thirteenth century water pump.

The land for the new airport used to be just sea, what led engineers to take on such an *arduous* task?

In the 1980s, Hong Kong was growing fast, the old airport Kai Tak was struggling to cope as passenger numbers rocketed. And Kai Tak Airport was famous for its perilous approach. The city needed a new airport. But where to build it? There was just no good and free space. The city is squeezed between the mountains on one side and the sea on the other. With space being such a rare and precious commodity, the engineers came up with a plan for the new airport that made everyone surprised.

Engineers look out to sea. They set about blasting and leveling two islands of the coast of Hong Kong ——Chek Lap Kok and Elm Check. The plan was to join them and create one big island out of the blasting rock, except the *seabed* between the islands was soft *marine* clay up to twenty meters deep.

8.2 Large-scale Reclamation & Land-making Projects by an Ancient Invention

The island needed a firm foundation, so they had to get rid of the clay. Engineer called the world's largest fleet of dredging ships to suck the soft clay all up. The *dredges* were like giant *vacuum* cleaners, and they rely on the connection that was invented from what is now Turkey eight hundred years ago. *Back then*, the ruler wanted to turn every land of a river into lush and green fields. An audacious genius inventor called Alzazeri took on the challenge of moving water up here. He created the world's first automatic *suction pump*.

Professor Alhasani is an expert on ancient technologies. He showed a replica of the ancient suction pump powered by a bicycle. The bicycle represents the waterwheel driven by the river, which provides the water to be pumped up the hill to distribute towns and villages. The piston is the heart of the machine, just like a squirt suck in water through a pipe from the pond, then drives it out through the *chamber* and up the outlet *hose* at the top. The flat valves mean water can only go in one direction——Upwards. When disturbing the soil in the river bed, water as well as the mud would be sucked up. Then the dredging started here. The 13th Century invention could also be useful for the Hong Kong in late 20th Century. Just like the suction pump can suck water, the dredging team of the Chek Lap Kok suck up *sludge* from the seabed. But the latter was conducted with faster speed——Ten tons a second for two years. With all the mud sucked up, engineers had a stable base on which to build. And by 1995, the island for Hong Kong new airport was complete.

8.3 The Greatest Challenge for Maritime Airports

Building Hong Kong International Airport in the South China Sea solved the problem of space, but airport in the sea would expose aircraft to some dangerous weather. Hong Kong International Airport sits in the pathway of deadly typhoons. But *meteorologists* can easily see a typhoon hundreds of kilometers wide as it races in. It was in fact an invisible threat to planes that airport designers were worried about.

The host: There is another kind of weather that can *ambush unwary* pilots, and it can be little for planes precisely because it is less predictable. It's *wind shear*.

Wind shear is any localized change in wind speed. And Hong Kong International Airport is located in a high risk zone between mountains and the ocean. Normal wind types don't apply. Rapid wind changes can affect the amount of lift on aircraft wings, causing planes to accelerate or stall *catastrophically*. Aircraft can literally fall out of the sky or be blown off the course. Fortunately, planes in Hong Kong are protected from wind shear. And incredibly, all is because of the connection to a brass band.

A scientist called Christopher Baisbey asked a brass band to help him test Doppler's revolutionary theory that you can detect something moving towards you or away from you by measuring change in frequency. It was over a hundred and fifty years ago, a brass band were used to *demonstrate* a scientific principle that we take for granted nowadays. But back then it had to be explained, demonstrated and proved. In 1845, the sciensist arranged the whole players on a train, to make the whole players moving very quickly towards and away from the scientist to finish this work. Today, with the technique improvement, we have substitutions to prove the theory, as shown in Fig. 8-1.

This model plane doesn't play tune, but it does make noise. And it moves fast. Listening carefully to the sound of the engine, as it gets closer and further away, the noise changes, but it doesn't change the volume to become louder or quieter. Instead, it changes in pitch, which is called the *Doppler effect*.

The Doppler effect works like this. The plane produces sound waves. When it travels towards us, the waves *bunch* closer together, changing frequency in creating a higher pitch sound. As the

Figure 8 – 1   The model plane to substitute the brass band

plane flies away, the sound waves stretch out, leading the frequency and pitch to change again. But what does the Doppler effect or Doppler shift have to do to detect the wind shear?

8.4　The Wind Shear Precaution by Radar and Light

The fast wind shear is potentially deadly invisible currents of air. By using Doppler's principle, Hong Kong *pioneered* an advanced wind shear warning system. The first line of *defense* for the airport is twelve kilometers from the runway. It is a long distance but because of the water, it has got a clear sight.

The radar detects wind shear by looking for water *droplets*, usually rain. The radar wave bounces off the droplets and back to the tower, as shown in Fig. 8 – 2. The airport can tell whether the rain is moving away from or toward the radar by measuring the change in frequency of the waves that bounces back, just like the sound of the model plane changed as it flew past.

Pilots can be alerted of big changes in frequency that can signal dangerous wind shear.

The host: Even at the bottom of the radar tower, there's no sense that this monotonous structure is watching, but it is doing its job twenty four hours a day, giving alerts to pilots flying into that Chek Lap Kok Airport.

But what if the water droplets are too small to detect or the air is too dry? In this situation, this Doppler radar is like being blind. So Hong Kong

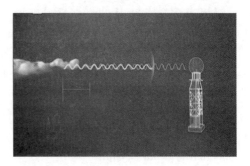

Figure 8 – 2   The radar detects the wind shear

International Airport went a stage further and pioneered a new way of using Doppler shift to detect the invisible threat. Anything that moves and makes a noise can produce the Doppler shift. That's why you can hear in police sirens, airplanes or a band on a train. But it's not limited to just noise. It could also detect the shift in another form, which is the light.

And a special kind of light is the key to Hong Kong's other use of Doppler. A small experiment was conducted as follows. The equipment was a fan and a small plane model. The giant fan was switched on, generating continuous current of air. But the airflow was invisible, which poses threats to the helicopter. When the small plane passed through the fan, the huge airflow blew it off its course and finally crashed.

In terms of the real airplane, the consequence would be catastrophic. The Hong Kong International Airport thought out a way to make the invisibility into visibility. Indeed the space around us is full of millions of tiny particles. They are not the dust that you might see in shaft of sunlight through a window. They are *microscopic* particles that you could fill hundreds of them across the width of a human hair. If you can see those moving, then you can see air current moving. The fan, the flour and a torch helped to demonstrate the fact.

The fan was switched on again, the flour was blown through the fan. Far way from the fan, there was nothing visible. But when the torch was turned on, the light was reflected by the moving particles and it could be seen. If the torch was turned off, nothing was visible again. The Hong Kong Airport adopted the similar principle by a more precise light of laser to detect the microscopic particles around. From the *unassuming* box on the top of airport, light is shot, as shown in Fig. 8 – 3. It is a laser called lidar.

Figure 8 – 3   The laser comes from this box at the airport roof

Hong Kong International Airport was the first to use Doppler in this way, keeping aircraft safe from wind shear. Engineers can take the box for getting planes in and out of Hong Kong new airport safely all thanks to a brass band.

8.5 The Light Roof without Pillars to Support

All of remains was to build one of the world's largest *enclosed* spaces for the new terminal. The designers' ambition for this new airport was very big that they wanted to create one of the biggest enclosed spaces in the world, and they wanted to be light and aerated at the same time.

A big space required a big roof. At seven hundred meters wide and over a kilometer long, this was not an ordinary roof. The airport *architects* insisted on a *jaw dropping* design, simulating the rolling waves of South China Sea. However for engineers, it was a *"nightmare"*. A roof in this size would need lots of muscular *columns* to take its weight. But the architect didn't want heavy weight structures, scattered up in light and spacious terminal.

The terminal is a huge space with nothing to stop passengers moving around freely. The *pil-*

lars have huge space among each other. So how does it realize that such a small number of slim columns to support a big roof? The answer lies in the light roof, but the light roof is not usually strong.

Interesting enough, making something light and strong is what plane designers have to do all the time. For the solution, the designers looked answers to the sky and found *aspiration* from a bomber of World War Two. During World War Two in Wellington, A twin-engine British bomber was used for nighttime raids over Germany. The crucial thing about the airframe was its strength.

British engineer Barney Wallis had found a revolutionary way of making a frame light and strong. It has more holes than metal, and the secret was in the way it arranged the material, as shown in Fig. 8-4. It's not about the amount of metal used, but how to use it. An experiment was conducted at the workshop to show how the less use of steel can function better. The technical advisor Martin Manning, as the chief structural engineer at Hong Kong International Airport was invited to take part in this experiment.

Figure 8-4 The material arrangement in the light frame

Firstly, a standard *I-beam* used in buildings all over the world was loaded up with 2.5 tons in weight, about the same weight as two families' cars. A chain hung the I-beam was shown in Fig. 8-5. As the chain slackened, the beam took the full load. The beam of 130 kilos' self-weight began to bend and it was barely enough to hold up.

Figure 8-5 The experiment setting for the I-beam

Then, the load was increased to 3.5 tons. At this time, the beam was unable to afford this load. It was seriously broken, as shown in Fig. 8 – 6. The steel failed completely at 3.5 tons. You could improve its strength by increasing its size, but that would also make it heavier. Amazingly, it was possible to send the I-beam "on a diet" as well as make it stronger.

Figure 8 – 6 The beam failed at the load of 3.5 tons

Try to think of the Wellington bomber, the answer lied in simple *geometry*. Taking a cue from Barney Wallis Wellington design, the steel was reorganized into a series of triangles. This *lattice* beam was almost twenty kilos less than the I-beam and much stronger if the calculation was accurate (Fig. 8 – 7). The load was spread through the lattice. So it won't bend. And the lattice beam would be tested under much higher load.

Figure 8 – 7 The triangle lattice beam

The lattice beam was loaded by all the bricks with a total weight of 4.5 tons. When the chain hung on the beam slackened, there was no change at all. The I-beam failed at 3.5 tons, which was estimated to be with 2.75 tons of *bearing capacity*. But the lattice beam was quite constant, which was almost twice capacity of the I-beam.

It shows that different arrangement of the steel makes a huge difference to the strength. A lattice was composed of very strong triangles, just like the Wellington bomber. And curved lattice of steel, also made of triangles, was exactly what the engineers needed to span the terminal roof in Hong Kong, as shown in Fig. 8 – 8. Each thirty six-meter span could be connected to just one huge but light lattice structure. This roof supports itself, without

Figure 8 – 8 The curved lattice roof in Hong Kong new airport

the need for hefty pillars getting in the way, leaving the space free for passengers to *roam around*.

8.6　Luggage Handling System

But of course, the building is just a shell for the business of passengers getting on the planes.

The host: Next, you need stuff inside that makes it work, like baggage handling. And the designers didn't want an ordinary system. They wanted one better than other airports.

Hong Kong International Airport carries forty million or so bags a year and 3.5 million tons of cargo. Keeping track of all the stuff is an immense task. When you check in for flight, you really want to be sure that your bags are going to come off the same airplane as you at the other end of your journey. Clearly, the designers at Hong Kong Airport realized just how much confidence matters the passengers. Airport around the globe rely on bar codes to identify a bag destination and owner, but bar codes have a success rate of only 85 percent. That means Hong Kong could lose six million bags one year, not counting half a million tons of *cargo* that will be a long queue at the lost luggage counter. The designers wanted to do better. So engineers had to pioneer a more advanced system.

They wanted technology that allowed bags to be scanned at a distance and with great accuracy. And they turned to a Cold War device for inspiration. In 1952 at Moscow, a security sweep revealed a bug at the American embassy, hidden in a replica of the *great seal* presented by the Russians over six years before. It was a revolutionary passive device operating without any kind of internal power supply. With no sign of the battery to power the bug, how was the signal being transmitted? The answer provided the solution to creating Hong Kong's new baggage handling system.

When a spy fired a specific radio frequency at the bug, it powered up the *microphone* and broadcast the *ambassador*'s secrets. A similar passive device classify luggage in Hong Kong. But how did it realize?

Well funny enough, spy bug technology is useful for sorting other things that don't know where they are going. At a farm deep in the British countryside, farmer Richard Weber manages his flock with a passive device, just like the spy device. An ear tag is a visual tag and a *microchip* is the heart of the whole device. When you switch it on, it will give you the automatic number. This device was fixed on lamb's ear to give a lamb an identity. When the ear tag was put in the ear, the ID number could be read by the *handheld* machine from the microchip, which is like the talks between handheld and the microchip. The information was put in the handheld including the gender, the type and other typical information. So from now on, whenever that ear tag was ignited, the microchip would be into action, triggering all information.

So the sheep have been tagged by a radio frequency identification chip or RFID tag for short. The sheep have been divided into three groups according to their value. Then, the sheep have been mixed up to find out if they could be back to their original groups by these microchips.

Yellow collars, red collars and collars with no colors were the identification of different groups. When a sheep ran through an auto draft, the auto draft would read the microchip to decide which pen it should go in.

The auto draft had doors to different sheep pens. Once it read the microchip from the sheep, it could open the corresponding door to the pen. The sheep sort machine sent a radio wave to activate the ear tag passive microchip, exactly like the spy bug. The ear tag replied its number and the computer checked its records of the sheep value, then it decided where to send the sheep in, regardless of the collar's color.

So the advantage of doing it this way is that it can be done at a distance. You don't have to be on site. They "talk" to each other automatically and sort with accuracy. So it's all about information and it is all about *readability*.

Hong Kong was one of the first airports to replace bar codes with RFID tags to track bags. Thanks to the spy bugs in Cold War, luggage can be sorted accurately at high speed with 97% of accuracy every twenty four hours getting to the right plane.

8.7 Wind Tunnel Tests

But the size of this building posed another problem for the airport engineers. Typhoons, as severe tropical storms, have their own international recognized scale of destruction. At the top of the scale, Category Five wind speeds can escalate to two hundred fifty kilometers an hour. These typhoons led *wreck havoc* in Hong Kong, sending everyone running. Hurricane force winds also caused problems for big buildings with high and flat sides, exactly like the Hong Kong Airport. To protect Hong Kong new airport from destruction, engineers knew the solution had to be strong enough to keep the building together. But just how strong did it have to be?

The pressure of moving air would be incredible. The walls could be reinforced to cope with the immense load. Unfortunately for the engineers, strong winds stressed the roof in a different way. Every airplane using this airport relies on one thing to get into the sky——Lift. One of the secrets to create a lift lies in air moving quickly across a surface. That terminal roof is a very big surface. As its fast moving air, there is high potential for a very big problem.

In the wind tunnel, the host was going to find out what happened to a curved roof when it was hit by a typhoon strength wind. A small building model and a curved roof similar with Hong Kong Airport's profile was built for the experiment as shown in Fig. 8-9.

The host got into the house for the test. This wind tunnel could reach wind speeds of a Category One Typhoon, namely 130km/h. The experi-

Figure 8-9 The roof and small house to test in the wind tunnel

ment started. At first, the roof was still in Fig. 8 – 10(a). With the wind speed increasing, the side wall was shaking. At the speed of 55km/h, all of a sudden, the roof "took off", as can be seen in Fig. 8 – 10(b). The wind force even broke the chains that linking roof and small house together.

Figure 8 – 10 The roof at different speeds

The host: You could just feel the lift, if it passed the critical point. The roof just wanted to take off, feeling like a wing. What I could feel that was the two stresses facing a building in a strong wind, because initially the wind was just hitting the side of the small house, and it wanted to *collapse* the building that way. When the wind grew stronger, suddenly there came a point when the roof was precarious, which was being lifted up by the strong wind. Thus, there were two extremely distinctive stresses acted on the building, one was to collapse the walls and the other was to lift the roof.

8.8 Flexible Connection between Roof & Wall

One way to hold a *flyaway* roof is to lock it down tight. But this would call for a *beefy* building, precisely what the architect didn't want. The airport engineers came up with a better idea, and the key was flexibility, thanks to a pioneering joint from the 1930's racing car.

The racing car called Silver Arrow, a 1930's Mercedes racing car, which featured a revolutionary new form of suspension that the Mercedes has been pioneering. *No sooner had* that suspension being pioneered, *than* it cascaded down from racing cars to more ordinary everyday cars you can still see in use today.

A *wishbone* is a triangle of strength and steel. It only allows movement up and down, stopping the wheels from moving in any other direction, as in Fig. 8 – 11.

Back in the workshop, a small demonstration was conducted to pit *flexibility* against stiffness. There is an old fable about a sturdy oak tree being blown down in the storm and a tiny aloe surviving because it could bend with the wind. A small illustration was devised to

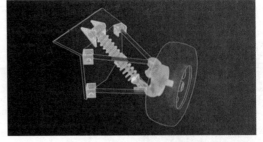

Figure 8 – 11 The wishbone

prove that flexibility can be a greater defense than stiffness. That was the solution at Hong Kong International Airport.

The suspension was *welded* to the first test car. There was no flexibility between the wheel and the car at all. The test was to lift the car and then drop it to the ground. As a result, the solid connection led the suspension coming through the bonnet and the car was totally ruined. So that was a *rigidity* approach.

Comparing with the rigidity, a flexible car with fully functioning suspension was under test. This time, the car was still flexible and it still worked to drive.

The car's wishbone inspired airport designers to join the roof and walls making a flexible joint allowing for movement during typhoons, as shown in Fig. 8 – 12. But unlike the car wishbones, airport engineers had to design for three directions of movements, not just one.

Figure 8 – 12 The wishbone used in the Hong Kong airport

The roof lifts like an aircraft wing, so the wishbone allows for up and down movement. But wind *blasting* the glass is transferred into a sideway's movement, so it also has a sliding *mechanism* for even greater flexibility and *hinge bearing* links wishbone to the slide bar, as shown in Fig. 8 – 13.

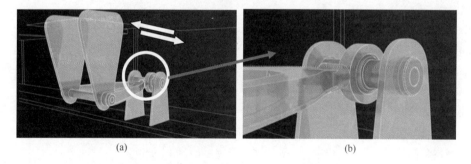

Figure 8 – 13 The sliding mechanism

A total of one thousand, three hundred wishbones allow the building to flex and move in typhoon winds, thanks to a *vintage* racing car.

As planes take to the sky, very few travelers will have any idea what an engineering achievement this airport really is, but it wouldn't have been possible without an eight-hundred-year-old water pump, a brass band, a cold war spy bug, a World War Two bomber, and a vintage racing car.

Words and Expressions

typhoon   n. 台风   bomber   n. 轰炸机

pump　n. 管道
vintage　adj. 古老的
reclamation　n. 开垦，收回，再利用
arduous　adj. 艰巨的
seabed　n. 海床
marine　n. 海运业
dredge　v. 疏浚（港口、河流等水域）
vacuum　adj. 真空的
chamber　n. 室，房间
hose　n. 软管
sludge　n. 污泥，烂泥
meteorologist　n. 气象学者
ambush　v. 埋伏
unwary　adj. 粗心的
catastrophically　adj. 灾难地
demonstrate vi. 演示，示范
bunch　v. 隆起，聚起
pioneer　vi. 倡导
defense　n. 防御措施
droplet　n. 液滴
microscopic　adj. 微观的
unassuming　adj. 谦逊的，低调的
enclosed　adj. 封闭的
architect　n. 建筑师
nightmare　n. 噩梦
column　n. 圆柱子
pillar　n. 柱子
aspiration　n. 抱负
I-beam　n. 工字梁
geometry　n. 几何学
lattice　n. 格子,格架

cargo　n. 货物
microphone　n. 扩音器
ambassador　n. 大使
microchip　n. 微芯片
handheld　adj. 手持型的
readability　n. 可读性
wreck　n. 破坏
havoc　n. 大破坏，浩劫
collapse　vt. 使倒塌
beefy　adj. 结实的
wishbone　n. 叉形杆
flexibility　n. 灵活性
weld　v. 焊接，使结合
rigidity　n. 硬度,刚性
blast vi. & vt.　爆炸
mechanism　n. 机制
brass band　铜管乐队
Chek Lap Kok Airport　赤腊角机场
back then　那时候，当时
suction pump　真空泵
wind shear　风切（变）
Doppler effect　多普勒效应
jaw dropping　令人瞠目结舌的，令人吃惊的
bearing capacity　承载力
roam around　散步,闲逛
great seal　国玺
wreck havoc　损坏
No sooner had…than…　一……就……
hinge bearing　铰接

Activities——Discussion, Speaking & Writing

1. Listening Practice

Watch the video clips, write down the key words you have heard, retell the main idea in your own words and then check your writing against the Chinese scripts below.

Exercise 1：机场的屋顶支撑

Chinese Script：但不同于汽车的双 A 臂悬吊，机场工程师在设计时必须兼顾三种移动方式，不只是一种而已。

Exercise 2：雷达监测风切
Chinese Script：通过测量反弹回的雷达波频率的变化，可以感知雨的移动是在远离还是在靠近，如同飞机飞过时，其声音变化一样。

Exercise 3：工字梁承载力试验
Chinese Script：工字梁是全球通用的建材，加在这根工字梁上的荷载重 2.5 吨，相当于两辆家庭房车的重量。一旦松开铁链，工字梁就会承受全部荷载，（铁链松开后）可以看出工字梁开始折弯了。这根重达 130 千克的工字梁仅勉强撑住了 2.5 吨的荷载。

Exercise 4：工字梁承载力
Chinses Script：当荷载加到 3.5 吨会怎么样？当锁链开始松开，工字梁再也撑不住了，那根工字梁严重变形，即工字梁在承载 3.5 吨的力时已经完全破坏。

Exercise 5：工字梁"瘦身"
Chinese Script：你可以把工字梁尺寸加大，来增加强度，但这也导致工字梁的重量增加。香港机场可不希望这样，谁也没想到我们既可以让工字梁瘦身，也可以增加强度。你想想惠灵顿的轰炸机吧。简单的几何学知识就能给我们提供答案。

Exercise 6：格构梁的重量
Chinese Script：如果计算没错，这根格构梁比工字梁轻了近 20 千克。它应该更加坚固，因为荷载平均分布在格构梁上，所以格构梁不会折弯。

Exercise 7：格构梁应用于航站楼屋顶
Chinese Script：钢材的排列方式会对其强度造成巨大影响。格构梁由非常坚固的三角形组成，就像惠灵顿轰炸机。而弧形的格构梁也是由三角形组成，这正是工程师所需要的，作为香港国际机场的航站楼屋顶。

Exercise 8：风洞试验
Chinese Script：现在风越来越大，而屋顶还稳如泰山。当风迎面吹来，整面墙都在振动。这时屋顶有些动静了，这表明是屋顶出了状况，屋顶开始晃动，而当风的时速刚过 55km/h 时，突然间屋顶就被掀开了。

2. Group-work：Presentation

Group (5 to 7 members) discuss and give your opinions about **Hong Kong's Ocean Airport.** Choose one theme below for your presentation.

Clearly deliver your points of the following themes to audiences. Each group collects information, searches for literature, pictures, etc. to support your points. Prepare some PowerPoint slides.

10 minutes per group (each member should cover your part at least one or two minutes).

NEED practice (individually and together)!! **Gesture** and **eye contact. Smile** is always Key!!

Themes(**Major topics included but not limited to**)

(1) Wind Resistance of Long Span Space and Suspension Structures
大跨空间与悬吊结构抗风
(2) Computational Wind Engineering Method and Application
计算风工程方法与应用
(3) Wind Tunnel and Its Test Technology
风洞及其试验技术
(4) Structural Wind Resistance Design Standard
结构抗风设计标准
(5) Structural Wind Disaster Risk Analysis and Assessment
结构风灾风险分析与评估
(6) Wind Engineering and Aerodynamic Problems
风工程和空气动力学问题
(7) Vibration Control and Intelligent Monitoring of Long-span Space Structures
大跨空间结构振动控制与智能监测
(8) Joint Connection of Space Grid Structures
空间网架结构的节点连接
(9) Benefits and Drawbacks of Reclamation
填海造地工程的好处与弊端
(10) Construction Technology of Land Reclamation Project
填海造地工程的施工技术

3. Independent work: Writing

Search for the famous **Ocean Airport of the world** in relevant index, like Engineering Index (EI), Scientific Citation Index (SCI) and China National Knowledge Infrastructure (CNKI). Then write a report independently on **the "Green" Design of Hong Kong's Ocean Airport.**

Unit 9

Engineering against Earthquakes

Teaching Guidance

Read the article, pay attention to the **Words and Expressions** and related **sentences** and **paragraphs**.

The world call of new lessons grows year by year, as earthquakes have shown, the loss of life from *collapsing buildings* is enormous. Designing and constructing to *withstand* earthquakes has never been more important. *Ironically*, the solutions may lie in the collapsed buildings, once the rescue team start to leave the ruin, it is the turn of global earthquake engineers. They come to try to unlock the very *puzzles* in each disaster, they are looking for the *clues* that someday may lead them to their *ultimate* goal—the earthquake-proof building.

9.1 Learning from New Events & Disasters

Earthquakes lead to death and *destruction*. But across the world, when engineers are fighting against the terrible earthquake disaster, there are solutions for buildings. There are solutions for industries, but above all, there are solutions to protect life. A *simulation* of the earthquake is not enough. Earthquake engineers over the world know that designing a safe building means studying a real earthquake.

Steve Shekerlian (Earthquake Engineer): The only efficient way is full scale tests, that means to study by real earthquakes.

Prof. K. C. Tsai (Earthquake Engineer): Earthquake engineering is actually a learning process, you know by experiencing new events and new damages.

Ron Egucchi (Earthquake Engineer): Unfortunately, earthquakes and mother nature, at this point, is probably the best sort of test site. We do a lot of tests in *laboratory* to see how buildings and many things perform, but this isn't until you get out, until you feel that you actually have the major earthquake that you will make sure what has worked and what hasn't worked.

In 1999, it was a busy year for the earthquake engineers. The first big earthquake struck in August, a 7.4 *magnitude* earthquake in a densely populated area Malmara in Turkey. In 45 seconds, 20,000 buildings were destroyed and 17,000 people lost their lives. Trying to conduct research in this sort of situation isn't like any other engineering job, it can be a painful task.

Ron Egucchi: When we got to Turkey, probably for my own personal perspective, what was very disturbing was that there were many buildings damaged, and knowing that there were still many people in the buildings. It was something very hard to take.

The first thing engineers must do is to carry up post *disaster review*. But entering the *aftermath* of the earthquake can be like arriving into a war zone. Devastated neighborhoods are no longer recognizable, and the earthquake engineers can't always figure out where they are. In the Turkey earthquake, Ron and his colleagues had to use Global Positioning System and *satellite map* to help them find out *specific* collapses that needed investigation. It doesn't take high-tech tools for engineers to know that a building built of cheap materials can *disintegrate* in seconds, no matter how well designed. Earthquakes have shown that unreinforced *concrete* is the worst sort of material to withstand an earthquake. In high-rise buildings, it will cause columns and beams to fail and storeys to collapse one on another, a *phenomenon* known as "Pan Caking" (Fig. 9 – 1).

Figure 9 – 1   The collapse form of Pan Caking

However simple or complex the failure is, there remains the puzzle of why one side of the street is a wipeout, while the other is hardly *scratched*. Engineers know that to understand what happens to buildings, they must start from the ground *beneath* their feet. Earthquakes travel through the ground in the form of P (short of primary) and S (short of secondary) waves. S waves are more *violent*, and it is what causes the damage to buildings. The type of ground through which they travel will *amplify* or decrease their *intensity*. The softer the soil is, the greater they are shaking. Just like sound and light waves, *seismic waves* can be reflected and focused, this can make damage in some neighborhoods worse but unpredictable. If a building *straddles* a *fault*, it may be ripped out. If it is constructed in certain kinds of soil, it may sink into *apparently* solid earth, because of a dangerous and dramatic process called "*liquefaction*". That is what happened to the lower floors of this Turkish apartment blocks. An extraordinary scene was the only time that liquefaction has ever been caught in a film, which was taken during 1963 Niigata Earthquake in Japan. When the earthquake happened, water poured up from the earth during shaking. This process has *tilt* buildings and caused them to *sink*. The most dramatic example of ever witnessed to USA took place in San Francisco *marine district* during the 1989 Loma Pulita Earthquake. Science teacher Ken Finn was leaving work when he noticed himself trapped into the parking lot which was *fluctuated* by soil liquefaction.

Ken Finn: It was a little bit up to 5 o'clock, and I was in my car getting ready to go home. When the car started to move up and down like someone was jumping on the *bumper*, I realized there was an earthquake, the ground was moving so much that I was hard to walk. It was to be very careful when I moved around. The parking lot fluctuated in waves about 3 feet up and down and in about 6 feet square regions. They were like standing waves, throughout the whole parking lot.

In the ground beneath Ken's feet, the earth was soft and had high water content close to the *surface*. When the shaking started, the loose soil *particles* separated from the surrounding liquid, then the heavier particles sank, water rose to the top, turning the ground to wet jelly. Driving enormous piles deep into the ground is a solution where earthquake engineers have identified liquefaction-prone area.

9.2 Taiwan Earthquake on September 12

There are other reasons besides soil condition and poor materials for building's collapsing. Earthquake engineers dream of catching an earthquake, recording enough information to know exactly what happens to different types of buildings during violent shakings.

In September 1999, their "dream" came true, when the year's second *massive* earthquake hit the heavily populated island of Taiwan in China. The earthquake was recorded by the most *sophisticated* seismic *instruments* and countless computers across the province. The information it gave engineers took them a step closer to understand other key reasons why buildings collapse.

Taiwan Province is one of the most seismically active places in the world, sitting on the earth surface with the seabed of Pacific Ocean, pushing them towards the mainland of Asia. When these pushing forces were released, the heavily industrial island was thrown into chaos. Before the earthquake, Puli was an attractive and popular tour destination in the center of Taiwan. Unfortunately, all that was about to change. Mrs. Qu with her husband and son live in a small apartment in the center of Puli Town. On the day before the earthquake, she was happily going about her business, everything was normal, except for one thing.

Mrs. Qu: Before the earthquake, the fish would jump around in our fish tank. Suddenly they began to strike the glass. Some were bleeding and some were even *fainted*. We just thought the fish were being foolish, and didn't realize a catastrophe was about to happen.

At 2:00 a.m., on September 21, 1999, a massive earthquake majoring 7.6 on the Richter scale occurred in Taiwan. The *epicenter* of the earthquake was at the tiny village called Jiji, Puli was dangerously close and suffered massive damage.

Mrs. Qu: That night at about two o'clock, the room began to shake up and down, from left to right. We heard noise, it was the glass smashing. My husband went to the child's bedroom to wake him up. I had a flashlight in my hand since there was no electricity. When we went downstairs to the *court yard* and found the *security room*, which was the only exit of the building that had completely collapsed.

Others didn't have such a lucky escape and needed *urgent* medical treatment. But in Puli, they were about to discover that one of the main hospitals had collapsed. Earthquake engineers are always learning from the terrible experiences and real events. In Taiwan earthquake, a *cruel* lesson was learned by one hospital. Hospital is an important building in earthquake disasters, meaning that they need not only to remain standing, but also to stay operational in an earthquake. But in Puli, the hospital had become one of the casualties, leaving only the parking lot as an ER (Emergency room).

Doctor Chang (Puli Veterans' Hospital): After the earthquake, the patients must be evacuated. Those could not walk had to be carried down from the third floor by our backs and put in the square in front of the hospital.

All the hospital's medical supplies have been trapped under the *rubble*. There was no electricity or water. Doctors were forced to do operations in darkness with only basic supplies, many victims died as a result. 100 miles north, in Taipei, Sun Qifeng was sleeping in a 12-storey apartment when the massive earthquake hit. The building collapsed in seconds, which was one of the few seriously damage buildings that were far from the epicenter. For Qifeng, it was the beginning of the life and death.

Qifeng: We felt the house started to shake and we thought it might not be anything too serious. And it would stop after a while. However, it shook stronger and stronger. We were scared and got up to try to escape. I opened the door and tried to get out. But it was too late. Everything above me fell down like that the earth had cracked and the sky exploded. I could not see anything. You can imagine sitting in a spaceship, closing your eyes and suddenly dropped to the earth. It was that sort of feeling at that time.

In the space of few seconds, the entire building collapsed, over 60 people lost their lives, including several of Qifeng's family members. *Incredibly*, despite jumping twelve floors and having several upper floors collapsing on the top of them, Qifeng and his younger brother were still alive, but they were trapped together at a tiny space, virtually unable to move. The boys talked to encourage each other in the darkness.

Qifeng: A couple of days later, we were talking when I looked at my watch and remembered that it was my birthday. My brother said to me, 'Qifeng, I never gave you anything on your birthday, and I have nothing to give you now. But I give you my favorite necklace as a birthday gift.' He then took off the necklace and put it around my neck. I was so moved.

Qifeng and his brother had buried themselves in the rubble for several days, surviving only by rotten fruit and dropping water. At night, they could hear the voice of rescuers, but no one had heard their despaired cries. By the six day, the two brothers had all given up, starving and delirious, they resigned themselves to death. But incredibly, digging in the rubble, Qifeng's younger brother managed to find a tiny hole and escaped to alert the rescue team to his brother's situation. Amazingly, both boys recovered physically from the experience, but the building collapse was something that they can never forget.

Qifeng: The fear is always there and we will never forget that experience. I remembered during the first month after the earthquake, my brother and I found it difficult to fall asleep. After we went to bed, we stared at the ceiling for fear that it would fall down again. All we wanted to do was to find a place to hide.

Back in Puli, the aftermath of the earthquake was terrible, even for those who were escaped. For weeks, nothing functioned.

Mrs. Qu: I thought Puli was completely destroyed. When I saw the buildings destroyed and people had died, I just wanted to cry. Every time when I came to my belongings, tears fell down. I thought that all of the families what they had built up and their whole lives were destroyed just because of the earthquake.

9.3 Different Earthquake Damage Modes

With the rescue operation was over, it was time for the earthquake engineers to see what could be learned from the *tragedy*, and patterns were starting to emerge. Professor Tsai from the Earthquake Engineering Research Center in Taipei has come to Puli to look at how *residents* were learning from the past and building for the future. He started his investigation at the hospital, and *revealed* why it didn't stay intact while other buildings surrounded were unharmed.

Prof. K. C. Tsai (Taipei Univ ersity): I think first of all, the shaking was very severe, and secondly, it is unlike the shorter buildings nearby, this building is taller, and in addition to that this building has more open spaces.

Hospitals need more open spaces for *wards* and *operating theaters*, but in an earthquake, open spaces could be a building's worst enemy, it means there are less walls to *stiffen* the structure.

Prof. K. C. Tsai: Relatively speaking, this building is weaker and softer, that's why this building was severely damaged.

The hospital engineers surrounded the columns with steel jackets, installing *reinforcing bars* to increase strength. Modifying the building after it has built is called "*retrofit*".

Prof. K. C. Tsai: After the retrofitting of the building, I think this building is much stronger now.

Elsewhere in Puli, Prof. Tsai saw positive steps being taken to try to prevent the same sort of damage from occurring again.

Prof. K. C. Tsai: In the old time, because it was difficult to afford good materials, people used simple *reinforced concrete* and bricks to build the buildings. Now after the Jiji earthquake, because the serious damage observed during the earthquake, people now are more willing to pay more to buy *steel columns* and steel beams for their buildings now.

Fractured columns and beams and poor reinforcement are common to many earthquakes. But the open spaces that weaken the hospital are common in traditional Taiwan buildings, causing a wide-spread and dramatical form of collapse. It was a phenomenon called "soft storey failure", and it grabs the attention of engineers around the world. A remarkable form of failure was shown as the complete flatting of the ground floor towards the space with merely meters high, but the upper floor could have hardly a window broken, as shown in Fig. 9 – 2. But how does this happen?

Prof. K. C. Tsai: For example, this building doesn't have walls in the direction parallel to the street direction, which makes the building relatively weak and soft during the earthquake. Especially if you have typical floors above the ground floor, you have regular partitions and regular walls in the same direction. It makes the building irregular on the ground floor, if we have a shaking in the direction parallel to the street, the building can easily fall down in this direction.

In addition to soft storey collapse, another pattern emerged. Its more striking *feature* was the height of the buildings which sustain most damage, the majority of which would be between eight and twelve floors. This was because the frequency of the earthquake was identical to the buildings, engineers call this *resonance*.

Figure 9-2　The ground floor collapses with the upper floor unharmed

　　But the new knowledge learned from earthquakes in Taiwan and other districts has helped engineers predict how all sorts of buildings will behave, and crucially, how they can design them to protect lives.

　　However, the experience in California is not optimistic. Although life goes on as normal, California lives in the *consistent shadow* of big earthquake. In Los Angeles school, there are usual earthquake drills for children. The children will climb down under the tables as soon as possible when they hear signals. But even though Californians know that earthquakes are inevitable, they don't always prepare for it. As a professor of architecture of the university in California, Berkeley, Mary Comerio has spent years researching Californians' attitudes towards major earthquakes. What she has found was not good news.

　　Prof. Mary Comerio: People in California don't remember the reality of earthquakes very much, although we live in this threat on a daily basis, there is a 70% chance that a major earthquake will occur somewhere in the Bay Area in the next 30 years. And somehow that probable risk doesn't translate into people's everyday life. In people's everyday life, they worry about whether they can pick up their kids on time from school, whether they can get their work done, whether they can meet their deadline. In their attitudes, *priority* will be given to dealing with *immediate* problems rather than preparing for the future.

9.4　Earthquake Precautions in California

　　The evidences from events like Malmara in Turkey and Taiwan in China have helped develop an amazing new software that can predict loss, an ordinary laptop computer can tell the building owner with what will happen if their building was hit by an earthquake. Fig. 9-3 shows the interface of the software.

　　Steve is a California Earthquake Engineer. In his work, he can use this technology to predict the earthquake effect at any address in America, like this one for example, a typical Californian building.

Steve Shekerlian: Today we have the technology in the capabilities of determining what level of damages we expect in buildings. And there are many computer programs that can do this, where you can basically input the data where the building is located, what type of construction it is, the age of the building. When we run a typical building, we can identify the relativity to where the hazards are. The process on the software is simple. We input the data, and we find out the location of the building, now what we do is to input every different structural *attributes*, and then we save that data, finally what we do is that we will put in the earthquake information. For example, we will load the da-

Figure 9-3 The interface of the software to predict loss

tabase of all the faults for five-hundred-year probability appearance. After that, computers will make hundreds of simulations based on disaster data, and you can identify the potential loss of this type of building during earthquakes, it also provides you the percent of damage and the dollar loss associated with it for different confidence levels.

But knowing what damage might be isn't the same as making precautions for it. Many people in high risk areas chose to do nothing about quake proofing, leaving their families in potential danger from the eventual and inevitable earthquakes. The most dramatic collapse in an earthquake maybe multi-storey apartment blocks or elevated freeways. But in fact, the biggest group of damage structures in the California quake is something much more down to earth——Single family *wood-frame* homes. Here at Los Angles, the city is trying to find a way to encourage the citizens to prepare themselves for a future quake. For the first time ever, the city is testing a system to grade homes according to their state of earthquake precautions. The better they protected, the better their rating is. Getting a poor mark could mean not being able to sell it out, while getting a good one could mean saving your family's life.

Shafat Qaze is a structural engineer. He is now carrying out an *assessment* on a traditional Californian home.

Shafat Qaze: What we are going to look here is for the *cavity wall*, the cavity wall is actually the *portion* between the foundation and floor truss. The 18-inch-high portion of this house typically carries the earthquake forces from the foundation to the house, and in the large earthquake, the foundation moves and the forces are transferred, we get a "scissor effect" (also means shear failure), the cavity wall will actually fail and the house will actually slide out. Going into the crawl space, it's kind of important to look for the *fixed mode* between the mudsill and the concrete foundation, we typically look for whether the house has actual bolts, because some of the old house do not have bolts. The bolts to connect the mudsill and the concrete foundation can help the house perform better in the earthquake. However, there is also area for improvement in this house. There is a typical concrete pile with a wooden pole sitting on top of it, supporting all the internal floor *trusses*. That typically is not good for a house, instead a continuous *timber struct* would be the best. But one good news is that it has a *diagonal bracing* that keeps it in position during the earthquake.

Most people in California wouldn't believe that they could save their house from disaster so simply or cheaply with a little retrofitting.

9.5 Effective Seismic Measures for General Engineering

However, not all retrofitting is cheap and easy. Steve Shekerlian works with major business owners in Los Angeles. In January 1994, he just completed a big retrofit for one of his clients——A major *brewery* who was gearing up for the Super Ball at their busiest time of year. Just as football players protect themselves from impact to come, Steve had to protect the North Ridge Brewery against the worst natural disaster——The earthquake. What he didn't realize was that his skills would be tested only in a matter of weeks. It was the biggest football match of the year just around the corner, damage could mean losing millions. Steve and his team surveyed the factory and decided to make some crucial changes, the main building was the first to be reinforced.

Steve Shekerlian: This is the *vintage* concrete shear wall structure of the 1950s, and earthquake force was resistant by the wall behind it. When we firstly analyzed the structure, many of these walls had openings for windows and ventilation openings. So what our analysis has determined was that some openings had to be closed up. For example, one solid section of wall was used to have a window similar to the left side. After filling it up, you have more solid walls where the loads can go down through the wall much more evenly.

Bracing was Steve's next weapon as shown in Fig. 9-4, from basic foundations to high-tech equipment, this simple form of strengthening seriously improve the chances of resisting earthquake forces, but Steve also had another *trick up his sleeves*.

Steve Shekerlian: This trick is called *seismic separation*. If two buildings are close to each other, during the earthquake, they will *vibrate*, they don't vibrate in exactly the same direction, sometimes they'll go towards each other or they'll go away from each other, and what this space does is to allow buildings to move without hitting the adjacent buildings and cause damages to the structure.

Figure 9-4 The bracing to defeat earthquakes

Steve's handy work would be tested within weeks, the biggest earthquake to hit this area in thirty years struck at the middle of the night. All over the North Ridge, buildings were shaking violently, it didn't look good for the brewery. Steve rushed to the North Ridge, but the first thing he thought about was his family.

Steve Shekerlian: My son and I got on the car immediately, because my parents lived in Mount Granada, very close to the epicenter, and I wanted to get up there as soon as possible to see how they were.

Steve's family had been lucky to escape unharmed, but elsewhere in the neighborhood, the damage was enormous.

Arriving at the factory, Steve discovered that the brewery has sustained *virtually* with no structural damage, and the company saved millions of dollars in lost *revenue*. The Super Ball watchers got their beer. It was a highly successful retrofit. But Steve had learned that being an earthquake engineer was more than just a job.

Steve Shekerlian: When you actually see it happens in your own backyard, and it was people that you know and building that you have designed. When the earthquake strikes, you are also affected, because you feel like you are part of them, and you want to make sure that nobody to be hurt in the buildings that you have designed.

Established technics have been very successful for conventional buildings, but large enterprises in the 21st century demand new solutions. Earthquake engineers can apply the most automatic and state-of-the-art technologies.

9.6 Earthquake Resistance in High-tech Industry Building

From Seattle to San Diego, from Taipei to Tokyo, many of the world's high-tech industries are located in the earthquake zones. These industries demand the highest possible level of seismic protection. In the 1999 earthquake in Taiwan, China, the high-tech Xinzhu Area as Taiwan's Silicon Valley was seriously affected. The companies here supplies one-third of the world's computer *chips* and the disruption could have a global implication. When the earthquake hit, the district was plunged into darkness when power lines *ruptured*. For the silicon foundries, loss of power meant lost of production. After two days of lost production for totally million dollars, the government has made an unprecedented decision to transfer the remaining power to Xinzhu. But power supply was not the only worries. These factories have many different expensive elements that needed protecting. John Dizon is the earthquake engineer responsible for the seismic performance of one of the biggest factories.

John Dizon: This is more complicated beyond life safety. Lots of facilities may run the year round, and business interruption is certainly one important factor. The company hopes to *resume* work immediately after the earthquake.

But for an earthquake engineer, the silicon chip factory is a real challenge. Pipes carry *poisonous* and highly *corrosive* gases, and must have emergency shut off bell linked to *seismographs*, to prevent serious disasters. Transportation is stopped whenever there is a small earthquake, and although the digonal bracing is needed to protect the building itself, when the building's contents are crucially important, they must be seismically protected too.

John Dizon: These devices are more expensive than construction and account for a large proportion of the total *assets* of the manufacturing plant.

In the clean room where the chips are actually produced, a single particle of dust can cause serious destruction, and the earthquake protection is to a much more exact standard.

John Dizon: In the clean room, we have many fixed systems to protect expensive equipment, like to prevent sliding or overturning in an earthquake.

However, if the contents of silicon chip plant are expensive, the contents of the world's research laboratories may be beyond price.

The University of California, Berkeley lies right on the top of Hayward fault, a recent seismic survey set alarm bells ringing, when they discovered the most valuable asset— "Vital New Research" was vulnerable to damage.

Prof. Mary Comerio: One of our surprising findings in our study of the Berkeley Campus was how highly concentrated our funded research is, about fifty percent of all funded researches are concentrated in seven buildings.

Like the silicon plants on Taiwan, Berkeley is a twenty-four-hour operation, and the equipment carries fabulous price tag. But if an earthquake hits here, the knowledge loss may be beyond price.

Prof. Mary Comerio: We really can't afford as major research *institution* to allow those buildings to be closed, we can't stop for two years, the cure for cancer may be lost if the *refrigerator* tips over with samples in it.

This is a new challenge for earthquake engineers in the future, and goes beyond the demands of the past, making sure that an earthquake causes virtually no damage to a specific building.

Prof. Mary Comerio: So, our goals are to figure out ways to continue operating and never shut the campus.

The solutions for Berkeley may lie in the untested, new technic of the cutting-edge earthquake engineering.

The ultimate goal for seismic engineers is the earthquake proof building. If you need your building to stand up in almost any earthquakes, there are many amazing solutions in the market, while some fantastic ones are on the *horizon*. *Base isolation* is a technic that literally separates the building from the ground beneath it. In a big earthquake, the ground will shake, but the building shouldn't. At that time, Los Angeles City Hall is the tallest building in the world ever to be base isolated. As an old building, it was an incredible challenge to retrofit.

Richard Puckowski: The most traditional way of *installing* an isolation system, is to build it from the ground towards up, but because the Los Angeles City Hall was built in 1928, we've got the difficulty of trying to install the *isolators* with the building in position, trying to minimize damage to this building.

When an earthquake hit, the city hall would be able to move four feet in any horizontal direction. In order for this to be possible, a 4-foot-wide trench was dug around the entire building.

Richard Puckowski: We have to leave room for the building to *occupy* a zone once the ground has removed away, as you can picture that the building standing vertically, and the ground moves, the building remains in place, it has to occupy space, so a trench was needed all around the city hall.

The next thing that had to be done was to separate the building from its existing foundations. To do this, firstly the engineers had to *excavate* down below the existing basement, and placed jacks under existing columns to keep the building erect. Next, 578 individual isolators would be placed between the existing foundations and the new basement, and the jacks would be removed to let the building rest on the isolators as shown in Fig. 9 – 5. Each isolator *consisted of interlayers* of rubber and steel, wrapped with a fire-proof shell. When the shaking starts, the bottom of the isolators move with the ground, but deforms to allow the building to stay almost stock still.

Richard Puckowski: The height of the basement is about 4 to 5 feet. The plane where the isolators install has isolated the entire building. When the earthquake strikes, the height below this elevation will shake with the ground. After the earthquake has completed, the foundation returns to its original state, the building presses to-

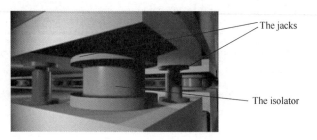

Figure 9 – 5 The isolator and the jacks

ward the bottom of the isolator, and it will eventually reach an *equilibrium* point and being *static*, all these occur within a few microseconds.

Another *cutting-edge* technology beginning to be increasingly important in building seismic protection is dampers. Damping technology began to be used in military applications about 20 years ago. It was designed to keep *intercontinental* missile operational despite the effect of near hit. Seismic engineers now use *dampers* to absorb the vibrations that will otherwise shake the building in an earthquake. Figure 9 – 6 shows the figure of a type of damper inside the building structure. Dampers are basically shock absorbers. This sort of dampers work by moving *fluid* from one end of the damper to the other through a small hole. Because of the force to *squeeze* the fluid through, the fluid becomes hot and the energy is transferred from movement into heat. When installed throughout a building, the damper absorbs resonance, reducing the shaking. But if the *molecules* of the fluid inside the damper could change their characteristics at command,

Figure 9 – 6 A damper in the building

engineers could make one part of the building *stiff*, while leaving others *flexible*, allowing the building to respond to the earthquake in the best possible way. This is the rudiment of the smart material.

Andrew Whittaker (University, at Buffalo): Smart materials for earthquake engineering applications are those materials whose characteristics can be changed in near *real-time*.

Engineers have already identified materials that can be changed in milliseconds by applying external forces.

Andrew Whittaker: The example can be shown as the liquid in the *injection* tubes as in Fig. 9-7. As we stroke the syringe backwards and forwards, we move fluid from one chamber to the other. When a magnet is put in the middle of the two tubes, it is hard to pull the liquid, and the properties of the liquid will change dramatically. This offers earthquake engineers unique opportunities in the future to improve how we can control the response of structures.

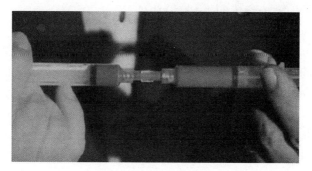

Figure 9-7 The liquids in the injection tubes

The key of that control may lie in powerful computers, together with smart materials. They could mean that *advent* of the truly intelligent building. But as we will see, the computer also allows engineers to think about other *out-range* solutions.

Seismic engineering has made rapid progress. With the knowledge obtained from recent earthquakes and advanced computer technology, engineers are getting closer to the goal of earthquake-resistant buildings. The future will hold any amazing possibilities.

9.7 Seismic Engineering Science & Technology in Future

Dr. George Lee is a leading *authority* on the seismic engineering in the future, a technology known as active control. He spent much of his career in biotechnology, applying engineer principles to the human body. Now, he wants buildings to learn from us. Above all, it is the human ability to balance themselves. Humans balance themselves by their *muscle system*, our brains and muscles work closely together to control our balance to an *extraordinary* degree. This is something that Dr. Lee sees around him every time he gets on the train to work.

Prof. George Lee(University, at Buffalo): We of course can see, can hear, can sense, we get the information to the brain, and the control tells the different muscles to act at different tension level, some totally relax, some pulls *intension*, some for example, you want to bend your knees that means you want to add some damping to minimize the impact.

Bending the knee is something that Tai Chi players take for granted, but it's actually a great achievement of human engineering.

Prof. George Lee: When I am taking on subways, I feel motions left and right. That is the *inertia force*, which is the weight timing the acceleration in the horizontal direction. This force wants to move my body off, so I must balance the horizontal force. I adjust three *variables*, the weight, the damping and the *stiffness*. If suddenly the car takes off, I'm leaving forward and backward, then I will balance myself in this direction.

Dr. Lee is working on a revolutionary prototype with this same system to be applied to buildings. Just as our *nerves* send the information to the brains, sensors placed around the buildings would feed data into computer, the computer would be linked to *hydraulic* and weight system which is *equivalent* with human's tendons and muscles.

Prof. George Lee: When the earthquake happens, the sensors measure the information about the acceleration and speed, feeding to the computer, and the brain decides that it is time to take action. For example, when activating *arrester* number one, it means to give more damping; Number two to give more stiffness; Number three to give partial stiffness, and so on. So you can get the best situation of balance and motion of the building against the ground motion.

The technology is known as active control, and it enables buildings to respond to earthquakes in real time. In buildings across districts in Japan and Taiwan Province of China, the first generation of the latest technology is being installed. A system known as *tuned mass damper* is at the cutting-edge of active control(Fig 9 – 8). During an earthquake, sensors throughout a building will send information to a central computer, this will activate a system of hydraulics that moves a counterweight, which is a huge metal block along a system of tracks. The engineers believe this should *counteract* the effects of resonance when the next quake arrives.

At that time, the tallest building in the world was under construction in Taipei, one of the world's most seismically active cities. The skyscraper combines the very best of conventional and cutting-edge seismic engineering. It would be supported by the biggest super columns in the building, and have the *skeleton* of dense steel reinforcing, as shown in Fig. 9 – 9. But with 88 floors above the ground, it would also have tuned mass dampers. No one would know how it would perform until the first earthquake hits.

Figure 9 – 8 The system of hydraulics and the track for counterweight

Figure 9 – 9 The building in Taipei under construction

Meanwhile, Dr. Lee also wants skyscrapers to learn how to fall over, just as we might protect our head when we fall. He wants them to behave *instinctively*, and protect the most valuable asset when they know they'll fall inevitably.

Prof. George Lee: The building actually has to fall in some way. When you want to protect some certain floors which save important equipment for example, you can conduct personalized *configuration* according to this to make something stable, something relax, and the rest to collapse at will.

Such dreams are a long way off, advanced technologies are incredibly expensive, and remains the preserve of the few. Most people effected by the earthquakes may never be able to afford them. Across the world, earthquake engineers each have a dream of something more *down to earth*.

Ron Egucchi: When asking most earthquake engineers what they wish, their answer may be that the technology and the design conditions will actually *implement* for all structures in the urban environment.

Others dream of the next generation being free from the fear of buildings that collapse in earthquakes.

Prof. K. C. Tsai: Earthquake doesn't kill people, it's the structure that made by men kills people. We sometimes have to think that we want to do something ahead of time, so that our children will not suffer the mega earthquake in the future.

But for all earthquake engineers, their mission remains the same.

Steve Shekerlian: We suppose to protect the public. When the public goes into the building that we have designed, they should feel safe and secure that the building is designed right and not going to fall down. As time goes on, we'll improve, and eventually, we will find a building that will be earthquake-resisting.

Words and Expressions

withstand v. 抵挡,反抗
ironically adv. 讽刺地,挖苦地
puzzle n. 难题
clue n. 线索,解药
ultimate adj. 最后的,最终的
destruction n. 摧毁,破坏
simulation n. 模拟,模仿
laboratory n. 实验室,研究室
magnitude n. 震级
aftermath n. 后果,余波
specific adj. 具体的,明确的
disintegrate v. 分解,瓦解,碎裂
concrete n. 混凝土
phenomenon n. 现象,事迹
scratch v. 刮伤,划伤
beneath prep. 在⋯下方

violent adj. 暴力的,剧烈的
amplify v. 增强,放大
intensity n. 强度,烈度
straddle v. 跨坐
fault n. 断层
apparently adv. 表面上
liquefaction n. 土壤液化
tilt v. 倾斜
sink v. 淹没,下沉
fluctuate v. 波动,涨落
bumper n. 保险杠
surface n. 表面,外观
particle n. 颗粒,粒子
massive adj. 大量的,大规模的
sophisticated adj. 复杂的,精密的
instrument n. 仪器,乐器

fainted　adj. 不省人事的,昏厥的
epicenter　n. 震中
urgent　adj. 紧急的,急迫的
cruel　adj. 残酷的,残忍的
rubble　n. 碎石,瓦砾
incredibly　adv. 难以置信地
tragedy　n. 悲剧,惨剧
resident　n. 居民
reveal　v. 揭露,泄露
ward　n. 病房
stiffen　v. 加固,加强
retrofit　n. & v. 补强
feature　n. 特征,特点
resonance　n. 共振,共鸣
priority　n. 优先,优先权
immediate　adj. 立即的,直接的
attribute　n. 属性
wood-frame　adj. 木结构的
assessment　n. 评估,评价
portion　n. 部分,一部分
truss　n. 桁架
brewery　n. 酿酒厂
vintage　adj. 古老的
vibrate　v. 振动,使振动
virtually　adv. 实际上
revenue　n. 收益,收入
chip　n. 芯片,晶片
rupture　v. 破裂
resume　v. 重新开始,继续
poisonous　adj. 有毒的,有害的
corrosive　adj. 腐蚀性
seismograph　n. 地震仪
asset　n. 财产,资产
institution　n. 机构,惯例
refrigerator　n. 冰箱
horizon　n. 地平线
install　v. 安装,装置
isolator　n. 隔离器
occupy　v. 占据,占有

excavate　v. 挖掘,发掘
interlayer　n. 夹层,隔层
equilibrium　n. 平衡,均势
static　adj. 静止的,静电的
cutting-edge　adj. 前沿的,最前沿的
intercontinental　adj. 洲际的
damper　n. 阻尼器
fluid　n. 液体
squeeze　v. 挤压,榨
molecule　n. 分子,微小颗粒
stiff　adj. 坚硬的
flexible　adj. 柔韧的
real-time　adj. 实时的
injection　n. 注射,注射剂
advent　n. 到来,出现
out-range　adj. 范围之外的
authority　n. 权威,权力
extraordinary　adj. 非凡的,特别的
intension　n. 张力
variable　n. 可变数
stiffness　n. 刚度
nerve　n. 神经
hydraulic　adj. 液压的
equivalent　adj. 相等的,相当的
arrester　n. 避雷器,制动器
counteract　v. 抵消,中和
skeleton　n. 骨骼
instinctively　adv. 本能地
configuration　n. 配置,设定
implement　v. 实施,执行
collapsing building　倒塌的建筑物
disaster review　灾难回顾
satellite map　卫星地图
seismic wave　地震波
marine area　沿海地区
court yard　庭院,中庭
security room　警卫室
operating theater　手术室
reinforcing bar　钢筋

reinforced concrete　钢筋混凝土
steel column　钢柱
consistent shadow　持续阴影
cavity wall　空心墙
fixed mode　固定方式
timber struct　木撑
diagonal bracing　斜撑
trick up one's sleeves　锦囊妙计

seismic separation　地震隔离
base isolation　基础隔震
consist of　由…组成,由…构成
muscle system　肌肉系统
inertia force　惯性力
tuned mass damper　调谐质量阻尼器
down to earth　实际的

Activities——Discussion, Speaking & Writing

1. Listening Practice

Watch the video clips, wirte down the key words you have heard, retell the main idea in your own words and then check your writing against the Chinese scripts below.

Exercise 1：千层派倒塌（连续倒塌）

Reference Script：高楼梁柱会因此倒塌，造成楼层坍塌，这一现象称为"千层派塌陷"（连续倒塌）。

Exercise 2：建筑隔震

Reference Script：在原有柱子下方放置千斤顶以支撑建筑，紧接着把 578 个独立的隔震器安装在原有地基和新的底层地下室之间，此时移除千斤顶，让建筑座落在隔震器上。每个隔震器由橡胶与钢铁间层组成，外面包上防火外壳。当地震来袭时，隔震器底部会和地面一起振动，隔震器的变形会让建筑几乎纹丝不动。

Exercise 3：阻尼器

Reference Script：阻尼器就是避震器。这种类型的阻尼器工作原理是让液体通过一个小孔从一端流到另一端，以达到吸震的目的。因为需要能量挤压液体，液体会变热，能量由动能转换成热能。阻尼器安装在建筑里能吸收共振，减少摇晃。

Exercise 4：台北大楼

Reference Script：台北正在盖世界上最高的大楼，这里是地震最活跃的城市之一，摩天大楼结合了传统技术与尖端地震工程技术，用最大的柱子支撑，并以密集的钢筋作为骨架，地上共 88 层楼也会安装调谐质量阻尼器。不过这栋大楼的抗震效果要等到首次地震时，才能验证。

Exercise 5：调谐主动控制

Reference Script：在地震来临时，感应器会把信息传给中央电脑，电脑会启动液压系统，在轨道上移动大金属块的配重。工程师相信此举可以消除共振的影响。

2. Group-work: Presentation

Group (5 to 7 members) discuss and give your opinions about **the Engineering against Earthquakes.** Choose one theme below for your presentation. Clearly deliver your points of the following themes to audiences. Each group collects information, searches for literature, pictures, etc. to support your points. Prepare some PowerPoint slides.

10 minutes per group (each member should cover your part at least one or two minutes).

NEED practice (individually and together)!! **Gesture** and **eye contact. Smile** is always Key!!

Themes(**Major topics included but not limited to**)

(1) Advanced Science and Technology of Seismic Engineering, Seismic Checking Calculation
地震工程的先进的科学技术，抗震验算
(2) Discussions on Seismic Engineering Practice
地震工程实践论述
(3) Strong Earthquake Observation and Analysis
强震观测与分析
(4) Seismic Risk Analysis of Civil Infrastructure
土木基础设施的地震危险性分析
(5) Site Effect and Geotechnical Seismic Engineering
场地效应和岩土地震工程
(6) Seismic Performance and Design Principle of Buildings and Lifeline Systems
建筑物与生命线系统的抗震性能和设计原理
(7) Application of Structural Control Technology and Performance Materials in Structural Response Control
结构控制技术和性能材料在结构反应控制中的应用
(8) Theory and Practice of Structural Health Diagnosis
结构健康诊断的理论和实践
(9) Codes and Standards for Seismic Design
抗震设计规范、标准
(10) Vibration Problems in Civil Buildings, Roads, Bridges and Tunnels
土木建筑、道路、桥梁、隧道等工程方面的振动问题

3. Independent work: Writing

Search for the famous **Engineering against Earthquakes** in relevant index, like Engineering Index (EI), Scientific Citation Index (SCI) and China National Knowledge Infrastructure (CNKI). Then write a report independently on **the Engineering against Earthquakes in China.**

Unit 10

Application of BIM in Shanghai Tower

Teaching Guidance

Read the text, pay attention to the **Words and Expressions** and related *sentences* and *paragraphs*.

Building information modeling (BIM) is not a specific software program. It is a *streamlined* process that allows us to make better decisions about project design based on reliable information analysis. The three core parts of BIM technology are cost control, optimization, and accurate construction. BIM's application value includes four aspects: visualization, parameterization, coordination, and integration.

10.1 What is BIM

BIM is an approach to the entire project life cycle, including design, construction, and *facilities management*. The BIM process supports the ability to *coordinate*, update, and share design data with team members *across disciplines*. The 3D process is aimed at achieving savings through *collaboration* and *visualization* of building *components* into an early design process that will *dictate* changes and *modifications* to the actual construction process. It helps engineers better predict a project's performance to increase safety, *constructability*, and *sustainability* before it is built, thus facilitating better decision making and more economic project delivery.

10.2 BIM throughout the Project Life-cycle

Use of BIM goes beyond the planning and design phase of the project, extending throughout the building life cycle, supporting processes including cost management, construction management, project management and facility operation.

1. Management of building information models

Building information models *span* the whole concept-to-occupation time-span. To ensure efficient management of information processes throughout this span, a BIM manager (also sometimes defined as a virtual design-to-construction, VDC, project manager – VDCPM) might be appointed. The BIM manager is retained by a design build team on the client's behalf from the pre-design phase onwards to develop and to track the object-oriented BIM against predicted and measured performance objectives, supporting *multi-disciplinary* building information models that drive analysis, schedules, take-off and logistics. Companies are also now considering developing BIMs in various levels of detail, since depending on the application of BIM, more or less detail is needed, and there is varying modeling effort associated with generating building information models at different levels of detail.

2. BIM in construction management

Participants in the building process are constantly challenged to deliver successful projects despite tight budgets, limited manpower, accelerated schedules, and limited or conflicting information. The significant disciplines such as architectural, structural and MEP designs should be well *coordinated*, as two things can't take place at the same place and time. Building Information Modeling aids in *collision* detection at the *initial* stage, identifying the exact location of *discrepancies*.

The BIM concept *envisages virtual* construction of a facility prior to its actual physical construction, in order to reduce uncertainty, improve safety, work out problems, and *simulate* and analyze potential impacts. Sub-contractors from every trade can input critical information into the model before beginning construction, with opportunities to *pre-fabricate* or *pre-assemble* some systems off-site. Waste can be *minimized* on-site and products delivered on a just-in-time basis rather than being stock-piled on-site.

Quantities and *shared properties of materials* can be *extracted* easily. Scopes of work can be *isolated* and defined. Systems, assemblies and sequences can be shown in a *relative scale* with the entire facility or group of facilities. BIM also prevents errors by enabling conflict or "clash detection" whereby the computer model visually highlights to the team where parts of the building (e.g., *structural frame* and building services pipes or ducts) may wrongly *intersect*.

3. BIM in facility operation

BIM can bridge the information loss associated with handling a project from design team, to construction team and to building owner/operator, by allowing each group to add to and reference back to all information they acquire during their period of contribution to the BIM model. This can yield benefits to the facility owner or operator.

For example, a building owner may find evidence of a leak in his building. Rather than exploring the physical building, he may turn to the model and see that a water valve is located in the suspect location. He could also have in the model the specific valve size, manufacturer, *part number*, and any other information ever researched in the past, pending adequate computing power. Such problems were initially addressed by Leite and Akinci when developing a *vulnerability representation* of facility contents and threats for supporting the identification of vulnerabilities in building emergencies.

Dynamic information about the building, such as sensor measurements and control signals from the building systems, can also be incorporated within BIM software to support analysis of building operation and *maintenance*.

There have been attempts at creating information models for older, *pre-existing* facilities. Approaches include referencing key metrics such as the Facility Condition Index(FCI), or using 3D laser-scanning surveys and photogrammetry techniques (both separately and in combination) to capture *accurate* measurements of the asset that can be used as the basis for a model. Trying to model a building constructed in, say 1927, requires numerous assumptions about design standards, building codes, construction methods, materials, etc., and is therefore more complex than building a model during design.

One of the challenges to the proper maintenance and management of existing facilities is understanding how BIM can be *utilized* to support a holistic understanding and implementation of building management practices and "cost of ownership" principles that support the full life cycle of a building. An American National Standard entitled APPA 1000 – Total Cost of Ownership for Facilities Asset Management incorporates BIM to factor in a variety of critical requirements and costs over the life-cycle of the building, including but not limited to: *replacement of energy*, utility, and safety systems; continual maintenance of the building exterior and interior and replacement of materials; updates to design and functionality; and *recapitalization* costs.

4. BIM in land administration and cadastre

BIM can potentially offer some benefit for managing stratified cadastral spaces in urban built environments. The first benefit would be enhancing visual communication of interweaved, stacked and complex cadastral spaces for non-specialists. The rich amount of *spatial* and semantic information about physical structures inside models can aid *comprehension* of cadastral boundaries, providing an *unambiguous delineation* of ownership, rights, responsibilities and restrictions. Additionally, using BIM to manage cadastral information could advance current land administration systems from a 2D-based and *analogue data environment* into a 3D digital, intelligent, interactive and dynamic one. BIM could also unlock value in the cadastral information by forming a bridge between that information and the interactive lifecycle and management of buildings.

10.3 Anticipated Future Potential

As a new technology, BIM has the following advantages: improved visualization; improved productivity due to easy *retrieval* of information; increased coordination of construction documents; *embedding* and linking of vital information such as vendors for specific materials, location of details and quantities required for estimation and tendering; increased speed of delivery; reduced costs.

BIM also contains most of the data needed for building performance analysis. The building properties in BIM can be used to automatically create the input file for building performance simulation and save a significant amount of time and effort. Moreover, automation of this process reduce errors and mismatches in the building performance simulation process.

Green Building XML(gbXML) is an emerging schema, a subset of the Building Information Modeling efforts, focused on green building design and operation. GbXML is used as input in several energy simulation engines. With the development of modern computer technology, a large number of building performance simulation tools are available. When choosing which simulation tool to use, the user must consider the tool's accuracy and reliability, considering the building information they have at hand, which will serve as input for the tool. Yezioro, Dong and Leite developed an artificial intelligence approach towards assessing building performance simulation results and found that more detailed simulation tools have the best simulation performance in terms of heating and cooling electricity consumption within 3% of mean absolute error.

10.4　Project Introduction of Shanghai Tower

A striking new addition to the Shanghai skyline is currently rising in the heart of the city's financial district. The super high-rise Shanghai Tower will soon stand as the world's second tallest building, and adjacent to two other iconic structures, the Jin Mao Tower and the Shanghai World Financial Center. The 121-story transparent glass tower will twist and taper as it rises, conveying a unique feeling of movement and growth, while reflecting the reemergence of Shanghai's economic and cultural influences amid the rise of an increasingly modern China. The massive mixed-use facility will include commercial and retail space; entertainment and cultural venues; a conference center; a luxury hotel; and public gardens, all evoking the sense of a self-contained city within Shanghai. A total construction area of about $574058m^2$, of which the building area of about $410139m^2$, underground construction area of $163919m^2$. The podium building is 32m high, the tower structure is 580m high, as shown in Fig. 10 – 1.

The 632-meter Shanghai Tower is the largest skyscraper in China as well as one of the most sustainable. The towering skyscraper comprises nine cylindrical buildings stacked on top of one another, all enclosed by a circular inner curtain wall and a triangular facade enveloping the entire structure. Each vertical neighborhood has its own atrium, featuring a public sky garden, together with cafes, restaurants, and retail space. The double-skinned facade creates a thermal buffer zone to minimize heat gain, and the spiraling nature of the outer facade

Figure 10 – 1　The Shanghai Tower

maximizes daylighting and views while reducing wind loads and conserving construction materials. To save energy, the facility includes its own wind farm and geothermal system. In addition, rainwater recovery and gray water recycling systems reduce water usage. The owner and design team are targeting a LEED Gold rating and a China 3 Star rating, ambitious goals for a project the size of the Shanghai Tower.

10.5　The Challenge of Shanghai Tower Project

Shanghai Tower is the tallest building under construction in Shanghai, and its equipment rooms are widely distributed on many sides. In addition to a large number of equipment rooms on the 1~5 underground floors, the number of above-ground equipment floors is as many as 20. Because the project has adopted several green environmental protection and energy saving technologies, it brings specific difficulties in projecting management and *system debugging*. The tower sets two energy centers in the low area and high area, which are divided into ten air conditioning zones including central refrigeration, ice storage, CCHP system, triple supply, ground source heat pump, VAV air conditioning and *fan coil*. With heat recovery device of fresh air and other systems, all those things make the whole system complex and increase requirements of the air-water system balance and automatic commissioning. Because the curtain wall is equipped with a radiator bracket and all projects set complex requirements, BIM modeling technology will be the best option for this project. BIM technology means to deepen the drawing and establish a three-dimensional model for conducting *pipeline collision detection* and comprehensive layout, form a prefabricated processing diagram in combination with *factory prefabrication*, and also it is used for labor planning and schedule control.

10.6 Overview of BIM Implementation

As we know, BIM technology has proven lately valuable efficiency in the systematization of different construction projects mainly in collaboration, coordination, and sharing data. It unifies the platform for all disciplines to share their design files. It is an effective information system that creates a database for all design and construction phases. This database considers the DNA of the building and can be a reference for future maintenance and development. For that reason, Shanghai's Tower Construction & Development decided to employ this technology in operating the design, structure, and construction processes of the tower. Jianping Gu, director and general manager of the company explains, "We knew that if we tried to work in a traditional way, using traditional tools and delivery systems, it would be extremely difficult to carry out this project successfully.

10.7 Geometric Design of Shanghai Tower

The flowing spiraling form of the tower was generated from a rounded triangular plan. This rounded triangle was derived from the relationship between the curved bank of Huangpu River, Jin Mao Tower, and the Shanghai World Financial Center. This attractive and distinctive form will be a milestone in Shanghai city for representing China as a global financial power.

Gensler Company designed the tower according to three main components that are parametrically modified and twisted.

1. Horizontal Profile

It is shaped as an equilateral triangle with smooth edges derived from two tangential curves, as shown in Fig. 10-2. There are two variables that shaped the profile: The radius of the large circle and its location to the center of the equilateral triangle.

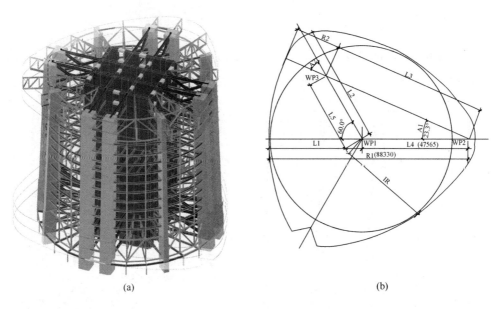

Figure 10 – 2   Horizontal Profile of Shanghai Tower

2. Vertical Profile

It is modeled by tapering the ground horizontal profile with the upper one, which resulted a right circular cone. The taper operation supported the function of the building. The wide lower profile afforded wide spans for the market and offices, whereas the slender upper plans provided short spans for the hotel, as in Fig. 10 – 3.

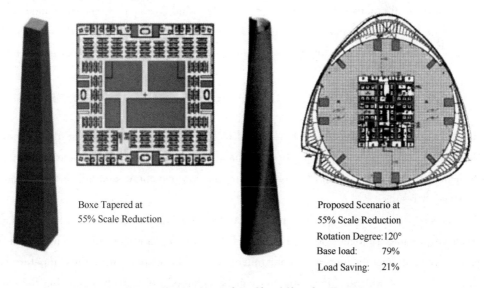

Figure 10 – 3   Vertical Profile of Shanghai Tower

3. Rate of Twist

It is a linear rotation operation from the base to the top. This process used different rotation angles to find out the best angle for the design, as in Fig. 10 – 4.

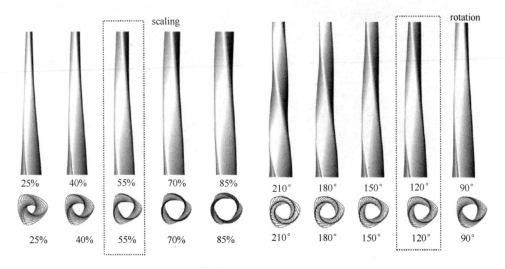

Figure 10 – 4 Rate of Twist of Shanghai Tower

The design team generated a number of alternatives for the geometry depending on two variables: the percentage of tapering operation and the angle of a twisting process. They built two models with different scales: 1: 85 and 1: 500. These models were examined by Wind Tunnel Test to determine the best alternative design that can bear the wind loads. The test concluded that the form should be tapered 55% and twisted 120°. These values create a geometry that reduces 24% in structural wind loads and cladding pressure and saves $ 50 million USD from structure budget.

10. 8 Comprehensive Pipeline Design in 3D Environment

In this project, the 3D visual design method of BIM is used to superimpose the professional models of architecture, structure and electromechanical engineering in the 3D environment, and import them into Autodesk Navisworks Software for collision detection, and adjust them according to the detection results.

In this way, not only the collision problem can be solved quickly, but also a more reasonable and beautiful pipeline arrangement can be created. In addition, through efficient field data management, immediate modification and rapid reflection into the model, an optimal pipeline layout plan highly consistent with the field situation can be obtained, effectively improving the success rate of a single installation and reducing rework, as in Fig. 10 – 5.

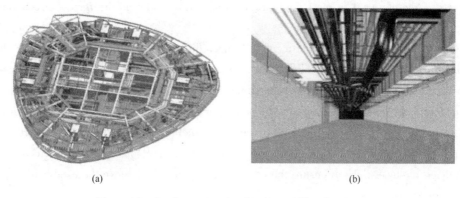

(a) (b)

Figure 10-5 Comprehensive Pipeline of Shanghai Tower

10.9 Use BIM to Arrange the Construction Schedule

For some large and long projects in the past, the traditional rough estimation method is mainly used in the preparation of schedule plan. In this project, the method of model statistics and simulation is used to arrange the construction schedule. Under the premise that there is no significant change in the total amount of the project and the total construction period, firstly relevant *parameters* (such as the shape parameters and performance parameters of all kinds of equipment, pipes, accessories, etc.) of the project amount statistics are added to the BIM model on the basis of the model in the deepening design stage.

Secondly, the engineering quantities of each section and system included in the model are classified and counted, so as to obtain the analysis of the engineering quantities of the subdivision and sub-system, and extract the data of equipment, materials and labor demand, etc.

Finally, based on the above data, the construction schedule is coordinated and arranged by considering the balance of various factors such as the delivery of work surface, supply of equipment and materials, labor resources, vertical transportation capacity, and use of temporary facilities.

By means of BIM model 4D, 5D function of statistics and simulation of the ability to change from past experience extensive, we can estimate the management pattern, and turn to a more scientific, more delicate, more balanced schedule arrangement method, in order to solve the construction peak of construction management confusion, lack of temporary facilities, lack of vertical transportation, and the contradiction of shortage of labor resources, and also avoid idle resources waste phenomenon such as equipment and facilities.

10.10 Use Models to Control Construction Quality

Since all collision points have been identified and solved one by one in the pipeline synthesis stage of the model, and the model is adjusted in real time according to the modification information on site, therefore, it is most appropriate to use BIM model as the *inspection* standard for construction according to drawings.

In this project, the project department will, according to the needs of the *supervision* department, import the image data of mechanical and electrical professional construction after completion into China for comparison. At the same time, the comparative results are analyzed and the "difference analysis report" is submitted. In particular, the problems affecting the system operation, elevation completion and subsequent process construction are recorded in the report in detail in the form of three-dimensional diagrams, providing basis for the supervision unit's next *rectification* and disposal suggestions to ensure the construction quality to achieve the established effect of the deepening design.

Words and Expressions

streamlined adj. 流线形的，效率更高的
coordinate v. 整合，协调
collaboration n. 合作，协作
visualization n. 可视化
component n. 成分，组件
dictate v. 支配，决定
modification n. 修改，修正
constructability n. 可施工性
sustainability n. 可持续性
span v. 跨度，贯穿
multi-disciplinary adj. 多学科的
coordinated adj. 协调的
collision n. 碰撞，冲突
initial adj. 最初的，最开始的
discrepancy n. 矛盾，差异
envisage v. 设想，想象
virtual adj. 虚拟的
simulate v. 模拟，模仿
pre-fabricate v. 预制，用预制构件组装
pre-assemble v. 预装配
minimize v. 最小化
extract v. 提取，提炼，获得
isolate v. 隔离
intersect v. 相交，交叉
maintenance n. 维修，维护
pre-existing adj. 预先存在的

accurate adj. 精确的
utilize v. 使用
recapitalization n. 资本重组
spatial adj. 空间的，空间上的
comprehension n. 理解
unambiguous adj. 清楚的，明确的
delineation n. 描绘，勾画
retrieval n. 检索
embed v. 嵌入，植入
parameter n. 参数
inspection n. 检查
rectification n. 改正，矫正
facility management 设施管理
across discipline 跨学科
shared properties of material 材料性质共享
relative scale 相对比例
structural frame 结构框架
part number 零件编号
vulnerability representation 脆弱性表现
replacement of energy 能源替代
analogue data environment 模拟数据环境
system debug 系统调试
fan coil 风机盘管
pipeline collision detection 管道碰撞检测
factory prefabrication 工厂预测

Activities——Discussion, Speaking & Writing

1. Presentation

Group (5 to 7 members) discuss and give your opinions about the *Development Trend of International BIM Technology*. Choose one theme for your presentation. Clearly deliver your points of the following themes to audiences. Each group collects information, searches for literature, pictures, etc. to support your points. Prepare some PowerPoint slides.

10 minutes per group (each member should cover your part at least one or two minutes).

NEED practice (individually and together)!! Gesture and eye contact. Smile is always Key!!

Themes (Major topics included but not limited to)

(1) Talent skills in BIM technology
　　BIM 技术的人才技能
(2) Development Trend of International BIM Technology
　　国际 BIM 技术发展趋势
(3) Use BIM to Design Green Buildings
　　应用 BIM 技术设计绿色建筑
(4) Use BIM to Comprehensive Pipeline Design
　　BIM 技术管道设计应用
(5) Use BIM to Arrange the Construction Schedule
　　BIM 技术安排施工进度计划
(6) Use BIM to Control Construction Quality
　　BIM 技术控制施工质量
(7) BIM Software Application and Green Building Innovation
　　BIM 软件应用与绿色建筑创新
(8) Building Industrialization and Assembly Architecture
　　建筑工业化和装配式建筑
(9) Development and Research of Green Building and Underground Comprehensive Corridor
　　绿色建筑和地下综合管廊的开发与研究

2. Writing

Search for the famous *Development Trend of International BIM Technology* in relevant index, like Engineering Index (EI), Scientific Citation Index (SCI) and China National Knowledge Infrastructure (CNKI). Then write a report independently on *the Practice and Future Potential of BIM*.

Unit 11
FIDIC Contract in Overseas Project Management

11.1 Brief Introduction of FIDIC

FIDIC is the French *acronym* for the International Federation of Consulting Engineers.

FIDIC was founded in 1913 by three national associations of consulting engineers within Europe. The objectives of forming the federation were to promote *in common* the professional interests of the member associations and to *disseminate* information of interest to members of its *component* national associations.

Today FIDIC membership numbers more than 60 countries from all parts of the globe and the *federation* represents most of the private practice consulting engineers in the world.

FIDIC arranges *seminars*, conferences and other events in the *furtherance* of its goals: maintenance of high ethical and professional standards; exchange of views and information; discussion of problems of *mutual* concern among member associations and representatives of the international financial *institutions*; and development of the consulting engineering industry in developing countries.

FIDIC publications include *proceedings* of various conferences and seminars, information for consulting engineers, project owners and international development agencies, standard pre-qualification forms, contract documents and client/consultant agreements. They are available from the *secretariat* in Switzerland.

11.2 Standard Forms and Applicability of FIDIC Contract

"***Conditions of Contract for Construction***", are recommended for building or engineering works designed by the Employer or by his representative, the Engineer. Under the usual *arrangements* for this type of contract, the Contractor constructs the works *in accordance with* a design provided by the Employer. However, the works may include some elements of Contractor-designed civil, mechanical, electrical and/or construction works.

"***Conditions of Contract for Plant and Design-Build***", are recommended for the *provision* of electrical and/or mechanical plant, and for the design and execution of building or engineering works. Under the usual arrangements for this type of contract, the Contractor designs and provides, in accordance with the Employer's requirements, plant and/or other works; which may include any combination of civil, mechanical, electrical and/or construction works.

"***Conditions of Contract for EPC/Turnkey Projects***", may be suitable for the provision on a *turnkey* basis of a process or power plant, of a factory or similar *facility* or of an infrastructure project or other type of development, where a) a higher degree of certainty of final price and time is required, and b) the Contractor takes total responsibility for the design and execution of the project, with little involvement of the Employer. Under the usual arrangements for turnkey projects, the Contractor carries out all the Engineering, Procurement and Construction (EPC), providing a fly-equipped facility, ready for operation (at the " turn of the key ").

"*Short Form of Contract*", which is recommended for building or engineering works of relatively small capital value. Depending on the type of work and the circumstances, this form may also be suitable for contracts of greater value, particularly for relatively simple or *repetitive* work or work of short duration. Under the usual arrangements for this type of contract, the Contractor constructs the works in accordance with a design provided by the Employer or by his representative (if any), but this form may also be suitable for a contract which includes, or *wholly comprises*, contractor-designed civil, mechanical, electrical and/or construction works.

The forms are recommended for general use where tenders are invited on an international basis. Modifications may be required in some *jurisdictions*, particularly if the conditions are to be used on domestic contracts. FIDIC considers the official and *authentic* texts to be the versions in the English language.

In the preparation of these Conditions of Contract for Construction, it was recognized that, while there are many sub-clauses which will be generally applicable, there are some sub-clauses which must necessarily vary to *take account of* the circumstances relevant to the particular contract. The sub-clauses which were considered to be applicable to many (but not all) contracts have been included in the General Conditions, in order to *facilitate* their *incorporation* into each contract.

The General Conditions and the Particular Conditions will together comprise the Conditions of Contract governing the rights and obligations of the parties. It will be necessary to prepare the Particular Conditions for each individual contract, and to take account of those sub-clauses in the General Conditions which mention the Particular Conditions.

11.3　Accurate Definition of FIDIC Contract Terminologies

"Contract" means the Contract Agreement, the Letter of Acceptance, the Letter of Tender, these Conditions, the Specification, the Drawings, the Schedules, and the further documents (if any) which are listed in the Contract Agreement or in the Letter of Acceptance.

"Letter of Acceptance" means the letter of formal acceptance, signed by the Employer, of the Letter of Tender, including any *annexed memoranda* comprising agreements between and signed by both Parties. If there is no such letter of acceptance, the expression "Letter of Acceptance" means the Contract Agreement and the date of issuing or receiving the Letter of Acceptance means the date of signing the Contract Agreement.

"Letter of Tender" means the document entitled letter of tender, which was completed by the Contractor and includes the signed offer to the Employer for the Works.

"Specification" means the document entitled specification, as included in the Contract, and any additions and modifications to the specification in accordance with the Contract. Such document specifies the Works.

"Drawings" means the drawings of the Works, as included in the Contract, and any additional and modified drawings issued by (or on behalf of) the Employer in accordance with the

Contract.

"Schedules" means the document(s) entitled schedules, completed by the Contractor and submitted with the Letter of Tender, as included in the Contract. Such document may include the Bill of Quantities, data, lists, and schedules of rates and/or prices.

"Tender" means the letter of Tender and all other documents which the Contractor submitted with the Letter of Tender, as included in the Contract.

"Appendix to Tender" means the completed pages entitled *appendix* to tender which are *appended* to and form part of the Letter of Tender.

"Bill of Quantities" and "Daywork Schedule" mean the documents so named (if any) which are comprised in the Schedules.

"Party" means the Employer or the Contractor, as the context requires.

"Employer" means the person named as employer in the Appendix to Tender and the legal *successors* in title to this person.

"Contractor" means the person(s) named as contractor in the Letter of Tender accepted by the Employer and the legal successors in title to this person(s).

"Subcontractor" means any person named in the Contract as a subcontractor, or any person appointed as a subcontractor, for a part of the Works; and the legal successors in title to each of these persons.

"Section" means a part of the Works specified in the Appendix to Tender as a Section.

"Contractor's Documents" means the calculations, computer programs and other software, drawings, manuals, models and other documents of a technical nature (if any) supplied by the Contractor under the Contract.

Performance Security

The Contractor shall obtain (at his cost) a Performance Security for proper performance, in the amount and currencies stated in the Appendix to Tender. The Contractor shall deliver the Performance Security to the Employer within 28 days after receiving the Letter of Acceptance and shall send a copy to the Engineer. The Performance Security shall be issued by an *entity* and from within a country (or other jurisdiction) approved by the Employer and shall be in the form annexed to the Particular Conditions or in another form approved by the Employer.

Variation

Variations may be *initiated* by the Engineer at any time prior to issuing the Taking-Over Certificate for the Works, either by an instruction or by a request for the Contractor to submit a *proposal*.

The Contractor shall execute and be bound by each Variation, unless the Contractor promptly gives notice to the Engineer stating (with supporting particulars) that the Contractor cannot *readily* obtain the Goods required for the Variation. Upon receiving this notice, the engineer shall cancel, confirm or vary the instruction.

Each Variation may include:

(a) changes to the quantities of any item of work included in the Contract (however such changes do not necessarily *constitute* a Variation);

(b) changes to the quality and other characteristics of any item of work;

(c) changes to the levels, positions and/or dimensions of any part of the Works;

(d) *omission* of any work unless it is to be carried out by others;

(e) any additional work, Plant, Materials or services necessary for the Permanent Works; including any associated Tests on Completion, *boreholes* and other testing and *exploratory* work;

(f) changes to the *sequence* or timing of the execution of the Works;

The Contractor shall not make any alteration and/or modification of the Permanent Works, unless and until the Engineer instructs or approves a Variation.

11.4 Documentary Management

File management is the carrier of management behavior and reflect the spirit of contract. File management of good quality will contribute to the better fluency and high efficiency of project management.

Priority of Documents

The documents forming the Contract are to be taken as *mutually explanatory* of one another. For the purposes of *interpretation*, the *priority* of the documents shall be in accordance with the following sequence:

(a) the Contract Agreement (if any);

(b) the Letter of Acceptance;

(c) the Letter of Tender;

(d) the Particular Conditions;

(e) these General Conditions;

(f) the Specification;

(g) the Drawings;

(h) the Schedules and any other documents forming part of the Contract.

If an *ambiguity* or *discrepancy* is found in the documents, the Engineer shall issue any necessary clarification or instruction.

Care and Supply of Documents

The Specification and Drawings shall be in the *custody* and care of the Employer. Unless otherwise stated in the Contract, two copies of the Contract and of each subsequent Drawing shall be supplied to the Contractor, who may make or request further copies at the cost of the Contractor.

Each of the Contractor's Documents shall be in the custody and care of the Contractor, unless and until taken over by the Employer. Unless otherwise stated in the Contract, the Contractor shall supply to the Engineer six copies of each of the Contractor's Documents.

The Contractor shall keep, on the Site, a copy of the Contract, publications named in the Specification, the Contractor's Documents (if any), the Drawings and Variations and other communications given under the Contract. The Employer's Personnel shall have the right of access to all these documents at all reasonable times.

If a party becomes aware of an error or defect of a technical nature in a document which was prepared for use in executing the Works, the Party shall promptly give notice to the other Party of such error or defect.

11.5　Time Bars under FIDIC Contract

Construction and engineering contracts often contain provisions specifying that, within a particular time, one party (traditionally the contractor) must notify the other (the employer and/or the contract administrator) of a claim or the likelihood that it might advance a claim. Sometimes these "time-bar" notice provisions are *elevated* beyond being merely an obligation, to the status of a condition *precedent* to being able to pursue a successful claim. If such provisions are *enforceable*, they can be severe: a failure to serve the required notice in the required *timescale* will be *fatal*, regardless of the merits of the underlying claim.

Unsurprisingly, many contractors have found themselves needing to construct *nuanced* arguments to get around the "trap" of clause 20.1. As the default final *dispute* resolution mechanism in FIDIC contracts is *arbitration*, reported court decisions giving guidance on the specific language used are rare, although some (but most definitely not all) other widely-used standard forms contain similar provisions. There are, nonetheless, a stock of standard arguments that are typically deployed.

Common law jurisdictions: the prevention principle

The prevention principle applies in the situation where one party has been prevented from fulfilling its contractual obligations by the conduct of the other. In such circumstances the second party cannot benefit from its own wrong by asserting the failure of the first. In conceptual terms relating this principle to time-bar provisions is attractive. There is a *manifest absurdity* (not to mention injustice) in an employer being able to benefit (via delay damages) from delaying its contractor if the contractor didn't go through the token exercise of notifying the employer about what the employer had done (or not done).

In the well-known case of Gaymark Investments v Walter Construction Group, the Australian courts held that, where the underlying delay was an act of employer prevention, the contract would need to contain an express right for the contract administrator to overlook a contractor's failure to *comply with* a time bar and extend the completion date regardless. This was thought to represent the clear position in Australia, and arguments based on this case were, for a time commonplace in England (and indeed elsewhere) too.

However, it was felt by many that under English law any hurdles presented by the prevention principle were overcome by the existence of an extension of time mechanism and that any time bar issue was separate. The English court's line seemed to be that if conditions precedent work smoothly in financing and other *transactions*, there should be no objection to layering such a mechanism on top of the right to bring a claim in construction contracts. Also, some *entitlements* to time and money arise not as a result of a default or a prevention, but simply as a function of the parties' agreed risk *allocation*.

The English court's position has been reasonably *consistent*. As long as the provision operated under the usual rules governing whether a condition precedent was valid, that's what it was. The Obrascon Huarte Lain SA v HM Attorney General for Gibraltar case in 2014 was widely reported and commented on, but it was more notable for containing a rare judicial analysis of the FIDIC wording rather than representing a departure from the now settled position on conditions precedent and the understanding that, *notwithstanding* the prevention principle, a well drafted condition precedent will apply.

The Scottish courts in City Inn Construction Ltd v Shepherd considered notions of *waiver* by the employer in the event that a claim was contemporaneously engaged with, notwithstanding a lack of contractor notices. There could be a valid argument around FIDIC sub-clause 20.1 based on waiver or *estoppel*, but those arguments tend to be very fact specific.

It remains to be seen whether the recent trend towards enforcing "no oral variation" clauses at the expense of inferring contract *amendments* through conduct (Rock Advertising Ltd v MWB Business Exchange Centres Ltd) might negate such arguments. Although the published FIDIC 1999 contracts do not contain "no oral variation" wording, it is often added in via the Particular Conditions.

Civil law jurisdictions: good faith arguments

In civil law jurisdictions (for instance in some of the GCC countries), an approximation of the prevention principle is often based on civil code concepts such as the obligation to perform one's obligations in good faith, and the *prohibitions* on abusing one's rights or being *unjustly* enriched. Each of these arguments can be attractive, but have limitations and/or uncertainties. It might sound like bad faith to rely on technicalities where one party's entitlements might be *extinguished*, but the duty of good faith doesn't rewrite a party's express contractual obligations. Also, many jurists consider that abuse of rights rules do not refer to contractual rights but to general legal rights, and therefore are not applicable when relying on a time-bar. As to unjust enrichment, if a contractual provision has been freely agreed by two parties and its operation happens to enrich one of them, is there anything necessarily unjust about that outcome?

Different, perhaps more robust, arguments arise because, typically, there is a prohibition on parties contracting out of mandatory prescription/limitation periods. The argument goes that it cannot be agreed that one party will lose its rights if it does not bring its claim within a shorter period than the one prescribed by law. Often these arguments do not distinguish between the enforcement of an existing entitlement and the conditions on an entitlement arising in the first place. There is no *definitive* view that they will succeed.

Solutions for *prudent* contractors

In general, a prudent contractor will seek to comply with the applicable notice provisions, rather than hanging its hat on iterations of the above arguments to get around a failure to comply with them. This makes sense under a FIDIC 1999-based arrangement, given that sub-clause 20.1

supplements other reporting obligations. For example there are requirements to issue monthly progress reports (sub-clause 4.21), which are actually supposed to refer to all the sub-clause 20.1 notices issued during the reporting period, and revised programmes to be issued whenever the previous programme is superseded by actual progress (sub-clause 8.3).

In the worst case, if these provisions have been followed, and the usual cycle of extra-contractual meetings have taken place and been well-minuted, it is usually possible to demonstrate that notice was given in any event – generally known as "constructive notice". This is especially the case as, despite its consequence, sub-clause 20.1 (even taking account of sub-clause 1.3 requirements) has little to say about the form and content of any such notice.

But although efforts are made to comply, it is fairly widely felt that clause 20.1 remains an onerous and draconian imposition on contractors. Employers' oft-cited justification for those provisions rings hollow. Most people involved in projects, or advising on them, will have heard stories of contractors being accused of being "overly contractual" for complying with notice provisions. In reality, *given the consequences of* not issuing them, it is surprising that contractors do not take a more scattergun approach to such notifications "just in case". The notion that these provisions promote good project management and prompt the early resolution and *mitigation* of issues in co-operation between the parties is often quite far from happens in reality on projects, and if that is the true justification, it seems surprising that sub-clauses 4.21 (Progress Reports) and 8.3 (Programme) were not given the same status.

FIDIC 2017: more notices

When FIDIC updated the Rainbow suite in 2017, *practitioners* were keen to see whether clause 20.1 would be retained, adjusted or deleted. The answer was perhaps surprising.

The time bar has been extended to apply to both parties equally. In itself this is not surprising, but early indications are that this has not gone down well with many employers ——even though clause 2.5 of the previous editions, regulating employer's claims, is widely thought also to act as a condition precedent——a view bolstered by the Privy Council in NH International (Caribbean) Ltd v National Insurance Property Development Company Ltd (Trinidad and Tobago).

The surprise comes in the new process, which fills a few pages of additional requirements. The engineer now has to serve a notice within 14 days if it thinks the claiming party has failed to serve its notice within the required 28 days. If the engineer fails to do that, the original contractor's notice will *be deemed to* have been valid. If the other party considers the notice should not be deemed valid even though the engineer failed to serve its notice, it can serve a further notice saying so —— the engineer will then need to determine the issues raised in the dispute about notices as well as the underlying claim. It then has to notify the parties accordingly. This has prompted more than one person to quip that the next FIDIC editions will probably require the contractor to serve a further notice if it wants to contend that the engineer's notice about the employer's notice about the engineer's lack of notice about the contractor's original notice was invalid.

FIDIC has also closed the *loophole* (if that's what it was) of constructive notice. Sub-clause

4.20 (the updated provision for progress reporting) has, tucked at the end, the following:

"However, nothing stated in any progress report shall constitute a Notice under a Sub-Clause of these Conditions".

It is hard to avoid the feeling this is a step away from pragmatism.

A radical proposal?

Some perspective might be helpful. On the one hand it seems fair and reasonable to impose the same notification obligations on the employer as the contractor. Furthermore, that balance might even have a more subtle effect – perhaps not wanting its own *compliance* with notification provisions to be put under the microscope, an employer might think twice before raising technical compliance issues against the contractor.

But, on the other hand, *heaping* notices upon notices (upon notices) does not seem like an *optimal* solution. For one thing it does not give the parties any additional clarity, but rather it enables more arguments and disputes about process rather than *substance* to arise, and provides more traps for the unwary (or those who are focused on executing the project).

Anyway, when it comes to formal disputes, employers rarely rely solely on technical time bar points when defending contractors' claims – it would be quite reckless to do so because tribunals will often look for ways around these technical arguments so as not to kybosh a *meritorious* claim. So it is not as though the notice provisions save anyone any time – they are usually a *distraction*.

Against that context, would it not be easier to remove the time-bar altogether? Failure to serve a required notice could be a breach of contract; and if the other party were able to demonstrate that some prejudice resulted from a failure to serve such notices, then that party could pursue the argument. But in many cases failure to serve a notice will be a purely nominal or technical issue (which should put the whole thing into context).

11.6 Contractor Shall Assume the Primary Liability for Delay

Progress management is an important aspect of project management. Following the time bars under FIDIC contract in last section, this section continues to discuss the causes and effects of delays. From the next paragraphs, we can conclude that contractor needs to assume the primary responsibility for delays. As a construction team, contractor should anticipate and take precautions against delays. As a developer, client should implement administration on contractor based on services and cooperate with contractor to prevent delays.

The problem of delays in the construction industry has always been a *ubiquitous* disease all over the world. In Saudi Arabia, Assaf and Al-Hejji found that only 30% of construction projects were completed within the scheduled completion dates and that the average time *overrun* was between 10% and 30%. In Nigeria, Ajanlekoko observed that the performance of the construction industry in terms of time was poor. Odeyinka and Yusif have shown that seven out of ten projects surveyed in Nigeria suffered delays in their execution. Ogunlana and Promkunton conducted a study on construction delays in Thailand. Al-Momani carried out a *quantitative* analysis on construction delays in Jordan. Frimpong et al. conducted a survey to identify and evaluate the relative importance of the significant factors contributing to delay and cost overruns in Ghana ground water construction projects. Chan and Kumaraswamy studied delays in Hong Kong, China construction industry. They emphasized that timely delivery of projects within budget and to the level of quality standard specified by the client is an index of successful project delivery.

Failure to achieve targeted time, budgeted cost and specified quality result in various unexpected negative effects on the projects. Normally, when the projects are delayed, they are either extended or *accelerated* and therefore, *incur* additional cost. The normal practices usually allow a percentage of the project cost as a *contingency allowance* in the contract price and this allowance is usually based on judgment. Although the contract parties agreed upon the extra time and cost *associated with* delay, in many cases there were problems between the owner and contractor as to whether the contractor *was entitled to* claim the extra cost. Such situations are usually involved questioning the facts, causal factors and contract interpretation. Therefore, delays in construction projects *give rise to* dissatisfaction to all the parties involved and the main role of the project manager is to make sure that the projects are completed within the budgeted time and cost.

The references mentioned in the last two paragraphs were issued more than 20 years ago and considered the causes or the effects of project delays separately. But they also indicated that the problem of delays in the construction industry has been *perplexing* us for a long time. On the other hand, construction sector is one of the important sectors that contribute to many country's economic growth, accounting for a considerable percentage of GDP and employing a large number of workers. The huge *volume* and *complexity* of projects pose a great challenge and provide a wealth of opportunities to various companies in the construction industry. Due to these points, it is necessary to take an *integrated* approach and attempts to analyze the impact of specific causes on specific effects and help the practitioners to prevent or *remedy* future delays.

We can identify major causes of delay and *categorize* them as client-related, contractor-related, consultant-related, material-related, labor-related, contract-related, contract relationship-related, and external factors. The major effects of delay can also be identified as time overrun, cost overrun, dispute, arbitration, *litigation*, and total *abandonment*. Identification of causes and effects alone does not help the project managers to take *appropriate remedial* or *preventive* steps. The project managers need to understand, for example, what causes or factors result in time overrun or cost overrun. Once these factors become clear, the managers can take *proactive* steps to avoid such situations. For example, if it is known that time overruns are *predominantly* caused by client-related factors, the project manager can: (a) make sure that payments for the completed work are paid on time, (b) reduce owner *interference*, (c) speed up the decision-making process, and (d) avoid unrealistic contract duration and requirements. Ten most important causes and six main effects of delay are as follows.

NO. 1 Cause: Contractor's Improper Planning

Local contractors often fail to come out with a practical and *workable* "work program" at the initial planning stage. This failure is *interrelated with* lack of *systematic* site management and inadequate contractor's experience towards the projects. The consultant only checks and reviews the work program submitted by the contractors based on experience and *intuitive* judgment. Improper planning at the initial stages of a project *manifests* throughout the project and causes delays at various stages. Only a project that is well planned can be well executed.

NO. 2 Cause: **Contractor's poor site management**

Contractor's poor site management is one of the most significant causes in causing the construction delays. Local contractors face *deficiency* in site planning, *implementation* and controls. A poor site management results in delays *in responding to* the issues that arise at the site and causes negative impact on the overall work progress.

NO. 3 Cause: **Inadequate contractor experience**

Odeh and Battaineh indicated that inadequate contractor experience was an important factor and this could be linked to the contract awarding procedure where most projects were awarded to the lowest bidder. A contractor with inadequate experience cannot plan and manage the projects properly and this can lead to *disastrous* consequences.

NO. 4 Cause: **Client's finance and payments for completed work**

Construction works involve huge amounts of money and most of the contractors find it very difficult to bear the heavy daily construction expenses when the payments are delayed. Work progress can be delayed due to the late payments from the clients because there is inadequate *cash flow* to support construction expenses especially for those contractors who are not financially *sound*.

NO. 5 Cause: **Problems with subcontractors**

Typically in huge projects, there are many subcontractors working under main contractors. If the subcontractor is *capable*, the project can be completed on time as planned. The project can be delayed if the subcontractor underperforms because of inadequate experience or capability. High degree of subcontracting leads to high risk of delays and this leads to *inefficiencies* in construction industry.

NO. 6 Cause: **Shortage in material**

Shortages in basic materials like sand, cement, stones, bricks, and iron can cause major delays in projects. Especially in the countries that are developing very fast, *oftentimes* demand exceeds the supply and this causes prices to increase. The contractors *postpone* the purchase activities until the prices decrease. Manavazhia and Adhikarib investigated material and equipment procurement delays in highway projects in Nepal and found these delays to cause cost overrun.

NO. 7 Cause: **Labor supply**

The quality and quantity of labor supply can have major impact on the projects. Take Malaysian construction industry as an example. About 20% of the workers are foreign workers, mainly from Indonesia and Vietnam. A few of them are illegal workers and their work quality is *relatively* low when compared to local laborers. The low quality and productivity of the foreign workers have impact on the project progress and efficiency. The illegal workers are frequently caught by the Malaysian immigrant officials and *deported* and this causes shortage of *labor pool* in the construction industry.

NO. 8 Cause: Equipment availability and failure

Many of the contractors do not own equipment that is required for the construction work. They rent the equipment when required. During the season when there are many construction projects, the equipment is in short supply and is poorly maintained. This leads to failure of the equipment causing the progress to be *hampered*.

NO. 9 Cause: Lack of communication between parties

Since there are many parties involved in a project (client, consultant, contractor, sub-contractors), the communication between the parties is very *crucial* for the success of the project. Proper communication channels between the various parties must be established during the planning stage. Any problem with communication can lead to severe misunderstanding and therefore, delays in the execution of the project.

NO. 10 Cause: Mistakes during the construction stage

The mistakes during the construction stage can be due to accidents, inadequate planning, or miscommunication between the parties. Whatever the reason, the mistakes can have impact on the progress of the project. While analyzing the effects of delays, there is no doubt that time and cost overruns can be ranked highly. Any delay in a project can lead to cost and time overruns and these two are linked. Whenever there are delays, there are disputes as to who should bear the responsibility and the cost. These disputes often lead to an arbitration process by third parties and failure in this process leads to litigation where the disputes are settled by the court. In extreme cases, some projects might be totally abandoned.

The delays can have negetive effects as shown below.

NO. 1 Effect: Time overrun

Client-related and contractor-related factors have impact on the time overrun. Out of the ten most important causes of delay discussed earlier, six causes belong to client-related and contractor-related factors. Factors such as in adequate planning by the contractors, improper site management by the contractors, inadequate project handling experience of contractors, and delay in the payments for the work completed directly affect the completion of the project and cause time overrun.

NO. 2 Effect: Cost overrun

Contract-related factors such as change orders (changes in the *deliverables* and requirements) and mistakes and discrepancies in the contract document result in cost overrun. Mistakes and discrepancies in the contract document can be in scope, deliverables, resources available and *allocated*, payment terms, achievement of various milestones, and the project duration. In most of the instances, time overrun leads to cost overrun.

NO. 3 Effect: Disputes

Client-related, contract-related, contract relationship-related, and external factors have impact on the disputes that arise during the course of the project. Factors such as delay in the payments for completed work, frequent owner interference, changing requirements, lack of communication between the various parties, problems with neighbors, and *unforeseen* site conditions give rise to disputes between the various parties. The disputes, if not resolved *amicably*, can lead to arbitration or litigation.

NO. 4 Effect: Arbitration

Client-related and contract relationship-related factors *escalate* disputes to be settled by arbitration process. A competent third-party can settle the disputes amicably without going to the court.

NO. 5 Effect: Litigation

Client-related, labor-related, contract-related, contract relationship-related, and external factors escalate disputes to be settled by the litigation process. The parties involved in the projects use litigation as a last *resort* to settle disputes.

NO. 6 Effect: Total abandonment

Client-related, consultant-related, labor-related, contract-related, and external factors contribute to the total abandonment of the projects. Take Malaysian construction industry as an example again, many projects were temporarily abandoned during the *financial crisis* between 1997 and 2000. Promoters of various projects *backed out* because of poor cash flow and economic conditions. Many of these projects have now become so *prohibitive* that they have been abandoned permanently.

Based on the discussions above, the next step is to provide solutions for clients, contractors, and consultants to reduce delays. They are divided into three groups: (a) solutions for the clients, (b) solutions for the consultants, and (c) solutions for the contractors. These solutions are expected to alleviate the problems faced by the construction industry.

Solutions for the clients

(a) While selecting the contractors, clients have to make sure that the contractors are not selected based only on the lowest bid. The selected contractor must have *sufficient* experience, technical capability, financial capability, and sufficient manpower to execute the project.

(b) Clients should not interfere frequently during the execution and keep making major changes to the requirements. This can cause *inordinate* delays in the project.

(c) Clients should have the finances in time to pay the contractors after completion of a work. Therefore, clients should work closely with the financing bodies and institutions to *release* the payment *on schedule*.

(d) Clients must make quick decisions to solve any problem that arise during the execution.

Solutions for the consultants

(a) While drawing the contract between the client and contractor, the consultant must include items such as duration of contract, mechanism to solve disputes, mechanism to *assess* the causes of delay.

(b) Consultants should prepare and approve drawings on time and monitor the work closely by making inspections at appropriate times.

Solutions for the contractors

（a）Contractors should not take up the job in which they do not have sufficient expertise.

（b）Contractors should have able site-managers for the smooth execution of work.

（c）Contractors must plan their work properly and provide the entire schedule to the clients.

（d）Contractors must make sure they have a sound financial backing.

Words and Expressions

acronym   n. 首字母缩略词
disseminate   v. 散布，传播
component   n. 组成部分，成分，组件，元件；adj. 组成的，构成的
federation   n. 联邦制国家，联合会
seminar   n. 研讨会，专题讨论会，研讨班
furtherance   n. 促进，助成，助长
mutual   adj. 相互的，共同的
institution   n. 机构，社会福利机构，习俗，制度，建立
proceeding   n. 诉讼，行动，会议记录
secretariat   n. 秘书处
arrangement   n. 安排，商定，布置，（乐曲的）改编
provision   n. 提供，供给，规定，准备
turnkey   adj. 交钥匙的，完整并可立即使用的
facility   n. 设施，附加服务，附加功能，天赋
repetitive   adj. 重复的，重复（因而乏味的）
wholly   adv. 完全地
comprise   vt. 包含，由…组成
jurisdiction   n. 司法权，管辖权，管辖范围
authentic   adj. 真实的，正宗的，地道的，逼真的，可靠的
facilitate   vt. 促进，帮助，使容易
incorporation   n. 法人，社团，公司，结合，合并，编入

annex   vt. 附加，获得，并吞
memorandum   n. 备忘录，简报
［pl.］memorandums / memoranda
appendix   n.（书、文件的）附录
append   vt. 附加，添加附件
successor   n. 继承者，继任者
entity   n. 实体，存在，本质
initiate   vt. 使开始，发起，使初步了解，传授，（常通过特殊仪式）使加入
proposal   n.（常为正式书面的）提议，求婚
readily   adv. 乐意地，容易地，迅速地
constitute   vt. 组成，构成
omission   n. 删节，遗漏，省略，排除
borehole   n. 钻孔，井眼，（为了探测石油或水）穿凿的深孔
exploratory   adj. 探索性的，探查的
sequence   n. 一系列，一连串，次序，顺序
mutually   adv. 相互地
explanatory   adj. 解释性的，说明的
interpretation   n. 解释，演奏，表演
priority   n. 优先，优先权，优先处理的事，优先次序
ambiguity   n. 不明确，含糊不清，模棱两可
discrepancy   n. 不符，矛盾，相差，差异
custody   n. 保管，抚养权，监护权，监护，拘留
elevate   vt. 提拔，抬高（地位），提高，

举起
precedent n. 先例；adj. 在前的，在先的
enforceable adj. （法律或协议）可实施的，可强制执行的
timescale n. 时段
fatal adj. 后果严重的，（事故、疾病）致命的
nuanced adj. 微妙的，具有细微差别的
dispute n. 争论，辩论，争端，纠纷；vt. 反驳，争夺（控制权、所有权）
arbitration n. 公断，仲裁
manifest adj. 明显的
absurdity n. 荒谬，谬论，荒谬的言行
transaction n. 交易，事务，办理，会报，学报
entitlement n. 权利，津贴
allocation n. （尤指经费）配置，分配决定
consistent adj. 始终如一的，（论点、观点）前后一致的
notwithstanding prep. & adv. 尽管
waiver n. 弃权，放弃，弃权者，弃权证书
estoppel n. 不容反悔，不许否认，禁止改口，禁止翻供
amendment n. 修正案，修改，改正，改善
prohibition n. （尤指通过法律的）禁止，阻止，禁令，禁律
unjustly adv. 不公正地，不公平地，不义地，不法地
extinguish vt. 使熄灭，使破灭，消除
definitive adj. 最后的，决定性的，不可更改的，确定的，权威性的
prudent adj. 谨慎的
supplement vt. 增加，补充；n. 补充，（报纸或杂志的）增刊
mitigation n. 减轻，缓和
practitioner n. 从业者

loophole n. 漏洞，枪眼，换气孔，射弹孔
compliance n. 遵从，顺从，服从
heap n. 一堆；vt. 堆放，大量地给予（赞扬、批评等）
optimal adj. 最优的，最佳的
substance n. 物质，实质，正确性，要旨
meritorious adj. 有功绩的，有价值的，值得称赞的，优异的，高素质的
distraction n. 分心的事物，注意力分散，消遣，心烦意乱
ubiquitous adj. 普遍存在的，无处不在的
overrun vt., vi. & n. （费用）超支；adj. 泛滥成灾的
quantitative adj. 数量的，与数量有关的，定量的
accelerate vt. 使…加快，使…增速；vi. 加速，促进，增加
incur vt. 招致，蒙受，遭受，引致，带来
contingency n. 可能发生的事；adj. 应变的
allowance n. 津贴，零用钱，允许，限额
perplex vt. 迷惑，使困惑
volume n. 量，容积，（书籍的）卷，合订本，音量
complexity n. 复杂，复杂性，复杂错综的事物
integrated adj. 各部分密切协调的，综合的，完整统一的
remedy vt. 补救，治疗，纠正，改善；n. 解决办法，治疗，疗法，药品
categorize vt. 把…分类
litigation n. 诉讼，起诉
abandonment n. 抛弃，中途放弃
appropriate adj. 适当的，恰当的，合适的
remedial adj. 补救的，纠正的，补习的，辅导的
preventive adj. 预防性的，防备的

proactive　adj. 积极行动的
predominantly　adv. 主要地，显著地，主导性地
interference　n. 干扰，冲突，干涉
workable　adj. 切实可行的，可经营的，能工作的
systematic　adj. 系统的，体系的，有计划的
intuitive　adj. 直觉的，凭直觉获知的
manifest　adj. 明显的，显现出的；vt. 显现出
deficiency　n. 缺乏，不足，缺点，缺陷
implementation　n. 实现，履行，安装启用
disastrous　adj. 灾难性的，损失惨重的，非常失败的
sound　adj. 有能力的，充足的，彻底的，熟睡的，资金充实的
capable　adj. 有能力的，有才能的
inefficiency　n. 效率低，无效率，无能
oftentimes　adv. 时常地；经常地
postpone　vt. 推迟
relatively　adv. 相对地
deport　vt. 把…驱逐出境
hamper　vt. 妨碍；n. 礼品大篮子，有盖大篮（尤装食品），洗衣篮
crucial　adj. 至关重要的，决定性的，定局的，决断的
deliverable　adj. 可以传送的，可交付使用的
allocate　vt. 分配

unforeseen　adj. 未预见到的，无法预料的，意料之外的
amicably　adv. 友好地，友善地
escalate　vt. 使…加剧
resort　vi. 不得不求助；n. 诉诸，度假胜地
prohibitive　adj. （费用）高得负担不起的，禁止的，禁止性的，抑制的
sufficient　adj. 足够的，充分的
inordinate　adj. 极度的，过度的，无节制的
release　vt. 释放，发射，让与，允许发表
assess　vt. 评估，计算，估算，评定，估价，对…征税
in common　共同的；共有的
in accordance with　按照，与…一致
take account of　考虑到，顾及，体谅
comply with　照做，遵守
given the consequences of　考虑到
be deemed to　被认为
associated with　与…有关系/联系
be entitled to　有权，有…的资格
give rise to　使…发生，引起
be interrelated with　与…相互关联
in responding to　回应，对…做出反应
cash flow　现金流
labor pool　劳动力储备
financial crisis　金融危机，财政危机
back out　退出，变卦，收回，食言，违约
on schedule　按时，按照预定时间

Further Reading
Reading Material 1

Working Together to Deliver a Brighter Future for Belt and Road Cooperation

Connectivity is vital to advancing Belt and Road cooperation. We need to promote a global partnership of connectivity to achieve common development and prosperity. I am confident that as we work closely together, we will transcend geographical distance and embark on a path of win-win cooperation.

共建"一带一路",关键是互联互通。我们应该构建全球互联互通伙伴关系,实现共同发展繁荣。我相信,只要大家齐心协力、守望相助,即使相隔万水千山,也一定能够走出一条互利共赢的康庄大道。

Infrastructure is the bedrock of connectivity, while the lack of infrastructure has held up the development of many countries. High-quality, sustainable, resilient, affordable, inclusive and accessible infrastructure projects can help countries fully leverage their resource endowment, better integrate into the global supply, industrial and value chains, and realize inter-connected development. To this end, China will continue to work with other parties to build a connectivity network centering on economic corridors such as the New Eurasian Land Bridge, supplemented by major transportation routes like the China-Europe Railway Express and the New International Land-Sea Trade Corridor and information expressway, and reinforced by major railway, port and pipeline projects. We will continue to make good use of the Belt and Road Special Lending Scheme, the Silk Road Fund, and various special investment funds, develop Silk Road theme bonds, and support the Multilateral Cooperation Center for Development Finance in its operation. We welcome the participation of multilateral and national financial institutions in BRI investment and financing and encourage third-market cooperation. With the involvement of multiple stakeholders, we can surely deliver benefits to all.

基础设施是互联互通的基石,也是许多国家发展面临的瓶颈。建设高质量、可持续、抗风险、价格合理、包容可及的基础设施,有利于各国充分发挥资源禀赋,更好融入全球供应链、产业链、价值链,实现联动发展。中国将同各方继续努力,构建以新亚欧大陆桥等经济走廊为引领,以中欧班列、陆海新通道等大通道和信息高速路为骨架,以铁路、港口、管网等为依托的互联互通网络。我们将继续发挥共建"一带一路"专项贷款、丝路基金、各类专项投资基金的作用,发展丝路主题债券,支持多边开发融资合作中心有效运作。我们欢迎多边和各国金融机构参与共建"一带一路"投融资,鼓励开展第三方市场合作,通过多方参与实现共同受益的目标。

——节选自国家主席习近平
在北京出席第二届"一带一路"国际合作高峰论坛开幕式发表的主旨演讲

Reading Material 2

CSCEC Works to Expand a Happy Living Environment

The following material is an image of China State Construction Engineering Corporation (short of CSCEC).

中国建筑，拓展幸福空间

When you draw the first line in your life, and take the first step of your lifetime, and when you come across someone and undergo something around you, you might not be aware that, at every moment of your long time, CSCEC, with its figure, keeps you accompanied all the time, just as a friend or relative.

  当你画下生命中的第一笔，
  迈出人生中的第一步，
  当你经历这个世间的每一个人，
  每一件事，
  你可能不知道，
  在你时间长河里的每一个瞬间，
  都有中国建筑的身影。
  像朋友，像亲人，
  始终陪伴在你身边，从未缺席。

From every single space below the ground, to every stretch of land you walk, and running across far-off deserts and distant mountains, or even extending the borders, CSCEC builds every road under your feet, and designs repeatedly the national territory from ancient to modern times.

  从地平线以下的空间，
  到你行走的每一片土地，
  甚至穿越千山万水，
  延伸至另一个国度，
  中国建筑铺垫你脚下的每一条路，
  也编织国家由古老走向现代的版图。

Looking up, you will see the height we constantly dream to break through. It is the great ambition in a city. There in each city, stand our masterpieces, which not only become urban landmarks, but also symbolize the temperament of the city. We are always the one, sharing every happy moment of your life. When you wish to return to nature, we are always ready to provide you with the most beautiful sceneries. Whenever you think of home, come to us, a habitat you can live in. Whatever life you choose, or wherever you want to go, CSCEC is always on your side as a true friend.

  当你抬头仰望，
  你知道那就是我们不断梦想打破的高度，是一座城市的雄心。
  在每一座城市，
  我们都留下自己的匠心之作。
  打造城市地标的同时，
  也影响着城市的气质。
  在你人生的每一个时刻，
  总有我们分享你的快乐。
  当你想回归自然和天性，
  我们也总是为你预留最美的风景。
  当你想到家的时候，
  其实，我们也是你的家的一部分。
  无论决定过什么样的生活，去什么样的地方，
  中国建筑，都如朋如好，与你相伴。

Under the Belt and Road Initiative,

CSCEC embraces an open approach to showcase in global market our understanding and practice of the happy living environment. Having built a great many significant projects in over 100 countries and regions around the world, CSCEC has integrated itself in world civilization, and in the least, helps shape the world.

在"一带一路"倡议的引领下,
中国建筑也以开放的姿态,
向全球市场传递着我们对幸福空间的理解和实践。
在全球100多个国家和地区承建诸多重要项目,积极融入世界文明的同时,
也以涓滴之力,
改变着世界的面貌。

Buildings originated out of the needs of human beings and should be people-oriented. Improving the living environment is the persistent pursuit of CSCEC. With the injection of frontier technology and green construction concept, CSCEC draws on its imagination and innovation capacity to provide people with the best living space. CSCEC has entered into global capital market by virtue of its super strength. An investment and construction group with the most internationally competitive power, is increasingly expanding the happy living environment for the people and the nation.

建筑因人而生,
以人为本,改善人居环境,
是中国建筑从未停止的追求。
将前沿科技和绿色环保理念完美注入每一个细节,
中国建筑以无穷的想象和创新能力,
致力于为人们提供最好的生活空间。
中国建筑也以超强的实力进军国际资本市场,
一个最具国际竞争力的投资建设集团,
正日益拓展着家国天下的幸福空间。

Hold fast to your persistent spirit, and share your heart touching feelings. We are bearing the responsibilities to get concerned at everyone's soul experience, and let their lives be cared for under all circumstances. CSCEC, we will always give you power, travel with you and be with you.

坚守你的坚守,
感动你的感动,
我们始终秉持责任和担当,
关切每个人的心灵体验,
让生命得到无所不在的关照。
中国建筑,给你力量,
伴你远行,与你同在。

中国建筑融媒体中心

To watch the film of CSCEC Image, please visit the official website at:
http:// english. CSCEC. com/About CSCEC/Companyprofile.

Appendixes, References & Acknowledgements

Appendix 1：Reference Translation for Parts of Paragraphs
精彩段落参考译文

Unit 2　伟大工程巡礼：纽约环保摩天楼

2.2　最佳岩石开挖方式

把库克的设计方案付诸实践并不是一件容易的事情。他不仅要在纽约建造最高的节能大楼，还要在最棘手的曼哈顿中城建造。在纽约打造摩天楼要面临巨大的挑战，特别是在第42街和第六大道交叉口，这是纽约最繁忙的十字路口之一。首先面临的困难是施工团队必须开挖曼哈顿中城最深的地基之一。

Serge Appel：这个地基有难以想象的深度，和整个工地一样宽，长200m、宽450m，深度大约100m。

为了完成这个地基，他们必须运走19万8千多立方米的废土。但让施工队头疼的是，曼哈顿是禁止爆破的。

Serge Appel：通俗来说，爆破是开挖地基最有效、最快速的方式，但第42街有地铁终端，另一条地铁路线来回行经第六大道，地铁站沿线不能进行爆破，附近还有不少历史建筑，旁边就有一栋50层高大楼。这里的地基开挖确实不适合用爆破的方式。

所以他们得采用传统方法，即开挖来解决问题。

Serge Appel：曼哈顿不能爆破，只能全靠大型挖掘机凿破和切削。

2.3　钢梁来自再生建材（废金属）

到2005年7月，终于打完地基。挖方工程总计近一年才完成。将近两年后，大厦盖到48楼，比周边街道高出200多米。工程继续如火如荼地进行，眼看还剩7层楼就盖完了。今天必须借助228米高的起重机把近36吨建材吊上大厦。工人在摩天楼脚下准备荷载物，起重机操作员处于太高的位置，以致他看不见数百米下的荷载物。这种情况下，操作员只能依靠大楼脚下施工人员的口头指挥吊起荷载物。

起重机的任务是吊起巨大的钢梁，这样工人才能继续建造剩下的楼层。但吊装之前首先需要解决的是把钢梁运到现场。这由装货平台的大拖车负责运输钢梁。

Thomas Kenney：这（一根钢梁）只是一般的荷载，大概重23吨左右。

但开这种拖车穿过车水马龙的市区，是一场噩梦。

Thomas Kenney：为避开车流，我们早晨5点就出发了。当Michael通知我们可以过来时，我们就上路了，但这一路上真是恐怖至极。到了中午，交通更加拥堵了。

Thomas Kenney：为了能够转弯，我需要开好长一段距离，我需要穿过三线车道，到那里才能左转，而问题是而这附近根本没有地方停车。

等钢梁一运到，工人就火速把20吨左右的钢梁吊上48楼。尽管看起来和普通钢梁没什么区别，但其实这是本大楼节能环保的秘密武器。

为了打造全球最具环保效益的摩天楼，湾布莱恩公园大厦从室内到室外的设计都将配备最尖端的绿色系统，建筑师理查库克的理念是彻底改变商业大楼的兴建方式。

Richard Cook：最根本的突破不是设计一栋大厦，让它"看起来"多么环保，而是从改变建筑的"DNA"入手，从根本上让我们重新思考建造大楼的方式。

库克的施工团队规定，许多兴建湾布莱恩公园大厦的建材必须从方圆800km内的场地取得，以降低交通费用和能源消耗。此外最基本的建材主要来自再生材料，这是库克的团队节省大楼能源的一种办法。兴建这栋摩天楼的所有钢梁至少有60%是再生建材，大楼的每根钢梁都来自当地一家专门收集废五金的工厂，由工人融化，废五金去除杂质并调整化学成分和材料结构。

Joe Stratman：首先，我们的钢梁原本是废五金，将近95%的废五金将作为原料来源，如旧车、旧洗衣机、旧洗碗机，我们绞碎、压缩、分类，基本上就可以开始制造钢梁了。

融化的液体流经模具，转化为一根根钢梁，接着用机器把钢梁推上冷却台，此时大多数钢梁仍因为高热而发光。一旦冷却之后，由工人测试强度和耐久性。这家炼钢厂的每根钢梁，最后运到曼哈顿的湾布莱恩公园大厦基地。不仅钢材可再生，就连炼钢厂的废弃物也可以制造可持续建材。炼钢厂会产生高炉矿渣，这种炼钢产生的附带废料通常直接丢弃，但湾布莱恩公园大厦的工地把废弃的矿渣回收利用，将作为混凝土的重要组成，占拌合料的45%，因而降低了混凝土生产过程中的碳排放。

2.4　高炉矿渣取代普通水泥

Richard Cook：当我们建造庞大的建筑时，小小的改变也会造成天壤之别。把普通的水泥换成高炉矿渣可以减少5万6千吨二氧化碳的产生，这个小小的改变真的实现了巨大的进步。

总计5万2千立方米的混凝土炉渣拌合料应用于超大结构，混凝土炉渣的使用不只降低了碳排放，还会增加混凝土的强度。以绿色建材兴建房屋本身就充满挑战，但湾布莱恩公园大厦施工团队同样也遇到了任何新建摩天大楼在施工时的困难，如无法预测的自然条件。在工地现场，滂沱大雨倾盆而下，雨水注满大楼回收系统，导致大多的工程施工被迫停工。

David Horowitz：大雨倾盆，施工平台湿滑，施工人员很难到平台上去，在一个安全的环境下继续安装钢梁，下雨天所有的钢梁安装和周边工作就到此为止。

但并非所有的施工项目都被大雨耽搁，为了赶上施工进度，有些施工项目必须继续施工。施工团队不畏风雨，在倾盆大雨中浇筑混凝土楼板。令人惊奇的是，大雨并没有影响混凝土的强度。混凝土工人需要保证和钢梁的安装同步进行，这样大楼才能准时竣工。随着大楼的升高，混凝土施工的进度也在加快。

第二天，天气放晴，阳光普照大地。今天的进度安排是把一根最大的钢梁吊上屋顶。而现在令人头疼的不是下雨，而是刮风。重达9千千克的巨大钢梁要被吊上268m的高度，它将作为屋顶尖塔的支承梁。施工团队以最严谨的预防措施，把钢梁绑在起重机上，希望避免任何严重的损坏。由于起重机无法横向控制荷载物，一旦刮风可能引起荷载物碰撞引发灾难。今天的风向不断改变。虽然团队很着急想动工，但此时也只能等待。他们为避免意外伤害而采取了一切可能的防护措施。过去 10 年，在美国仅起重机的意外每年会平均夺走 82 条人命。2006 年 11 月，在西雅图发生整个起重机倒塌事件，造成一人死亡，损坏了附近三栋大楼。事发前 8 个月，迈阿密有个起重机操作员坠地身亡，由于他脚下的安全平台破坏了起重机的系杆。类似的意外频频发生，所以起吊这根钢梁要做好充分的准备。

Michael Keen：因为风势太大，我们今天要把荷载物送回。我们要把横梁吊到近 800 尺位置，显然现在风势太大，我们担心荷载物会翻转撞上起重机，所以我们只能等明早再试。

（尽管困难重重），但他们今天必须盖完这座超大建筑的 52 层楼。说起来容易做起来难，施工现场离地面超过 243m，每个人都得爬上两座 12m 的梯子，穿过迷宫般的钢梁才能到达工作现场。在这种高度施工，工人须具有过人的胆识，因为这是没有安全网的"高空钢索表演"。起重机把钢梁运送过去，随后工人们必须把每根钢梁小心就位，像拼图似地扣在一起，需要胆大和心细，而工人们毫不畏惧。终于他们成功了，安装最后一根钢梁，他们完成了摩天楼的 52 层。每一天，一层又一层的建设让这个超大结构每天朝尖塔的安装更近一步。他们需要建造高 54 层楼的太厦，这之后尖塔才能安装完成。和许多摩天大楼的建造一样，大楼继续向上延伸，但这栋大楼的独特之处就在其绿色特征。不过这栋大楼并非史上第一栋环保摩天楼，施工团队参考学习了其他建筑师和工程师的技术。

2.8　未来绿色建筑的无限畅想

最后一根钢梁吊上顶楼，但工作尚未结束。这栋超大结构完成了 288m 的超级高度后，接着要迎接下一个重大挑战：在巨大钢结构的顶端安装一座 78m 高的尖塔。

等待了好几个星期，时机终于到来了，但施工人员刚开始安装，就发生了意外。

新闻播报：中城一座起重机的吊斗从 53 楼高处坠下，撞碎了施工中的美国银行大楼的几片窗户。

一个吊斗撞进大楼里，打碎了窗户，玻璃碎片和材料坠落至 245m 之下的街道。

路人 A：我听见了很大的声音，看到那些东西掉下来。

路人 B：吊斗撞上大厦，连同大楼的一些材料一起坠落，可能有些碎片掉落击中了大楼的窗户。

由于破碎的窗户紧邻起重机支架，紧急事故小组立即出动，确定起重机的结构是否仍坚固，并清除了破碎的玻璃。4 名行人和 4 名工作人员在这次事故中受伤。这次意外让工地之外方圆两个街区都遭到了封锁。所有工程完全停滞，施工团队重新整编。花了近两周时间处理完所有安全问题后，施工人员终于重新着手进行最后一件重要的作业，给湾布莱恩公园大厦加上顶冠。在钢构架顶端竖立一座巨大尖塔后，大楼的主要结构才真正到达总高度 365m。庞大的尖塔必须用拖车一块一块运来工地，把庞大荷载送进曼哈顿必须特别小心。

Michael Keen：进纽约市基本上会多花两天，因为必须趁晚上进城，接着一大早就得到工地卸货。人们说得没错，这是个不夜城，路上永远不乏车流。有时候难免危险，不过后面有陪同车保护。

第二天一大早，尖塔的组件终于运到工地。

尖塔由 70 件钢构件构成，需要一件件吊上屋顶，然后在屋顶上组装。如果天气理想的话，预计需要 5 星期左右。不过尽管起雾，施工仍照常进行。起重机把一个长 7.5m，至少重 1.5 吨的组件吊起。在近 275m 高处，稍有不慎便足以致命。工人小心把组件导引就位，把组件焊接和用螺栓连接。作业顺利进行。接下来几个星期，尖塔越盖越高，终于，湾布莱恩公园大厦实现了 365m 的高度。当大厦运营时，将成为全球最环保的大楼，并可能成为都市未来的领航者。这栋大楼预计可节省 50% 的能源和水的消耗，住户也会从中受益。

Anne Finucane：他们将拥有更健康的环境，呼吸更干净的空气。他们所处的环境，不只令他们骄傲，我们也会提供一个更好的物质环境，这是未来的美好蓝图。

可是绿色建筑的成本不菲，加上全球气温上升，湾布莱恩公园大厦是否具有影响力？

Stuart Gaffin：一栋大楼本身不会有什么影响，我们必须大力推广，因为曼哈顿是个小岛，我们必须改变大量的建筑表面、系统和科技，才能对气候发生影响。

很多人相信建筑界有感染力的风潮可能已经开始出现，而且能够引领市场。

Paul Goldberger：湾布莱恩公园大厦的许多观念已经融入其他建筑物，再过几年，我们会看到许多这样的建筑。

Richard Cook：我并不认为这个设计一点问题也没有，但现在正是个引爆点，也许我们已经跨过去，越过了那条界线。我们的目标是尽可能兴建最环保的建筑。因此大伙儿一边看一边想美国银行怎么会决定花重金建造一栋真正的环保大楼？我希望这栋大楼能推动全美国和全世界其他更多的建筑实现绿色设计。

但库克对湾布莱恩公园大厦的美好愿望仅仅是迈向重大改变的起步。摩天楼的绿色建筑不断演化。这并不是一个简单的过程，但这一步对我们的城市和地球未来的发展极为重要。至少就目前看来，绿色建筑有着无限的发展空间。

Unit 3　中国超大桥梁

3.2　建造桥拱的挑战

卢浦大桥于 2000 年 10 月动工，工人把两端支撑的桥桩打进去。拱桥有 64 个组件，钢铁工人正在打造第一个。这个设计极为复杂，由两个相同的向内弯的桥拱组成。根据组件位置的不同，每一个组件制造的规格都不一样。当组件运送到现场，每个组件由巨大的起重机缓慢地吊装定位。这种施工作业相当棘手，有些组件重达 15 万千克。每次工人装好一个新组件，起重机就从他们正上方往前滑过准备吊装下一个组件。随着桥拱越盖越高，工程师面临着如何保持桥拱垂直竖立的挑战。没有支撑，桥拱的两侧势必落水。但搭建脚手架会阻碍过往船只通行。造桥者的解决方案是由上往下固定桥拱：首先工人搭建临时铁塔，桥梁两端各有一座 120 多米的铁塔，他们用钢缆把桥拱系在铁塔上，每个桥拱组件由两根缆索支撑，桥拱一块一块朝中间越盖越高，附加的缆索防止铁塔倾倒，支架的工作量巨大，如同建造另外一座桥梁一般。这个如同斜拉桥的支撑方式会在拱桥完工后立即拆除。在桥拱完成前，这个组合支撑体系岌岌可危，桥拱每向前一步就更加脆弱。

"要让两端衔接起来,桥拱才能成为一个有效的结构体系。因为没有压力,两端不但会上下摇晃,也很容易受风势影响。"结构工程师 Mark Ketchum 说道。理论上造桥者必须尽快闭合桥拱,但实际施工进度非常缓慢,因为必须从内部焊接桥梁组件。内部空间非常狭窄,通风很差,焊接炬需要加热到 60℃。周围太热了以致焊接工人一次只能工作 15 分钟。最终完成需要焊接 300 多公里。而在外面,勘测员需要确保桥拱没有偏离既定位置。重力、温度和风力都对桥梁有很大的影响,会改变桥梁的尺寸和位置。一旦设计工程师计算错误,最后一段组件将无法匹配,意味着桥拱无法闭合。这样直接导致桥拱岌岌可危,因为台风随时可能侵袭。这一地区每年至少遭受两次台风袭击,强烈台风的风势足够轻易扭开脆弱的接点,吹垮支撑塔,以及导致缆索瞬间断裂。而光凭桥拱的重量,就足以让整座桥梁坠入河中。

在动工之前,卢浦大桥的设计师将桥梁模型置于 7 级地震的地震作用力和时速 270km/h 的台风下。测试结果显示桥拱可以通过大自然最严厉的考验,即便它尚未完工。但即使这样,桥拱没有闭合之前造桥者也无法高枕无忧。经费预算、施工进度以及整个施工项目的安全性,全看桥拱能否顺利密合,桥拱越大出错的空间也越大。

Mark Ketchum:凡是由构件一块一块组装建造出来的桥梁,都可能建造到正中央时发现最后一块组件无法闭合。

施工顺利进行数月之后,工程好像慢了下来,卢浦大桥的高级设计师解释是因为安装最后一片组件时,出了一点麻烦。大桥仍在建设过程中,由于尚未完工而变得不堪一击,特别是此刻毁灭性的台风即将来袭。

2002 年 7 月,动工将近 2 年之后,中国的卢浦大桥耸立在上海的天际线上。此刻工程还没竣工,最脆弱的时刻又碰上最糟糕的天气。台风"威马逊"在太平洋渐渐形成,距离上海只有 130km。暴风连续 24 小时袭击中国沿海,强风把树木吹倒,破坏房屋,不断吹袭着未完工的卢浦大桥。这场台风造成上海市 5 人死亡,随后台风转向北边消失了。卢浦大桥的支撑塔在这次台风中稳住了,桥梁毫发无伤。但造桥者知道下次未必能这么幸运,因此必须在下次台风来袭前实现桥拱闭合。工程师准备安装最后一块组件。最后一片组件吊装上去,但此时能否闭合仍是未知的。工人很快焊接好组件的一边,但另外一边的空隙太宽了,他们在等待外面的温度上升,好让钢材热胀使空隙密合。终于在 20℃实现了空隙密和,拧紧最后一个螺栓。钢桥拱施工终于大功告成。但巨大的桥拱尚未"脱离危险",工程师仍需采取措施防止桥拱向外塌陷。

3.6 固定主缆的关键技术

1980 年在佛罗里达州,2 万吨的尖峰冒险号撞毁了阳光航线桥。这艘船在恶劣气候下航行在佛州的坦帕湾,因雷达失灵偏离航道一头撞进桥墩导致中央跨倒塌。撞桥事件造成 35 人死亡,包括从桥上翻落的巴士所载运的 26 名乘客。润扬大桥的建设者不容许这种悲剧发生,他们会尽量让桥梁立足在陆地上。所有桥型中,只有一种能够跨越这么长的水上距离并让桥塔靠近河岸,那就是悬索桥。润扬大桥就像每一座现代悬索桥一样,依赖于两条从此岸延伸到彼岸的主缆。1m 厚的缆索两端牢固地固定在锚墩,但剩下的主缆却是活动的,包括桥塔顶端的缆索也是这样。桥塔就像帐篷的支柱一样把缆索撑起,但缆索仅仅是挂在桥塔顶上,而非固定在上面。如果把主缆锚固在桥塔顶端,桥塔无法承受缆索的拉力，会向内侧倾覆，导致整个结构失效。在主缆上吊着360条悬索承受桥跨的重量。这些悬索的连接点也是活动的，这种柔性连接可以帮助桥梁抵抗地震和台风的重击。在当时，完成后的润扬大桥将成为全球第三大吊桥，桥塔相当于70层高楼的高度，桥跨将长达几乎1.5km。只有一座桥塔坐落在长江的江面上。截至2001年，全球只有两条桥的跨距超越了这个长度，分别是丹麦的大贝尔特大桥和日本的明石海峡大桥。

2001年2月，润扬大桥动工了，这项工程对中国的发展意义非同寻常。这座大桥预计5年完工，第一步是建造桥塔，因为缆索向下传递的巨大力是由桥塔承担的。工程师要把桥塔固定在岩床上，但由于长江的淤泥层太厚，为了打造坚固的基脚，施工人员必须在每个桥塔基地沉陷32根桩柱。这些桩柱就像巨大的高跷竖立在下面的岩床上，支撑200m高的桥塔。要保证桥塔屹立不倒，这个施工步骤非常关键，万一桥塔开始倾斜将会造成严重的损失。除此之外，还有一个关键问题需要解决，即固定主缆的锚墩一定要稳如泰山。但这里的土壤泥泞松软，缆索可能把整个锚墩拖离原位。对于整个工程而言，克服这个弱点，将是整个建筑项目最困难、最危险的任务。

在当时，润扬大桥将成为世界第三大悬索桥。主缆的荷载巨大，其拖拉锚墩的力将到达6800万千克。设计师必须采取措施防止锚墩移动。为此造桥者必须把锚墩向地底延伸至少30m，造就了当时中国历史上最大的锚墩。可当工程师探测河岸时，发现土壤不只是泥泞而已，还富含丰富的地下水，这给施工带来了更大的困难。工程师必须设法引开地下水，否则开挖的坑洞随时可能灌进泥浆水，地下水的快速涌入，可能困住里面的施工人员。工程师想出一个解决方案，他们决定打造巨大的地下围堰，借此阻挡地下水流入。在3400m^2的基地周围，工人开挖了一系列的壕沟，并用钢筋混凝土填满。这些墙会防止地下水流入，让里面的开挖工作更安全。但是还有一个问题就是当时国内唯一的深沟挖掘机被江北的工人拿去使用了，江南的承包商不得不采用更冒险的做法，他们决定冻结土壤，来控制深邃的地下水。

因此他们必须建造一间大冷冻厂，工厂把盐水冷冻到零下20℃。盐水的冰点比淡水低很多。布置的管线把水循环到施工场地周围30多米以下。整个系统就像大冷冻库里的线圈。超冷的水把土壤冻结成1m厚的墙壁，可以阻挡地下水。没有流进的地下水，这两块基地的挖掘工作变得比较安全。挖掘的工作夜以继日地进行，工人想在夏天洪水淹没他们的挖掘基地前完工。每个基坑至少30m深，每隔4m就必须浇灌混凝土支撑架。工人挖得越深，地下水冲垮围堰的危险就越大。这是因为基坑越深，地下水压力就越大。

Manchung Tang：挖得越深，就要进行更多的冷冻作业。当坑洞挖到一定程度，水就会从底下非常快地冒上来。

开挖6个月以后，两个锚墩向下沉陷了至少9层楼高。它们看起来很像停车场，每个锚墩足以容纳7千辆车。数百名工人在两个场地上施工。季风季节及泛滥的洪水再过几星期就到了，这将会大大增加工程的危险性。在当时，兴建国内最长的悬索桥必须完成两座全国最大的锚墩，锚墩必须能够固定巨大的主缆，这是整座桥的安全保障。目前工人已经挖掘出28万m^3的泥土，但工程师们认为再挖下去就太冒险了，于是下令施工团队开始进入下一个阶段的施工项目。工人把两个锚墩填满31万m^3的混凝土和砂石。润扬大桥的工程师赢得了与滚滚长江抗争的第一场比赛。

3.8 桥梁抗风技术的关键

这些钢箱梁十分关键，因为直接决定润扬大桥能否运营数十年还是很快就倒塌。美国一架悬索桥的命运就是如此。1940 年 7 月，华盛顿州的塔科马海峡大桥通车，这是当时全球第三大悬索桥。这座桥竣工不久就发现问题，微风吹过就会引起桥梁晃动，因此这座桥被戏称为"舞动的格蒂"。4 个月后，时速仅 67 公里的强风，让桥梁扭曲起伏，桥梁越摇越厉害，并迅速恶化至毁灭性的"和谐振动"效应，过了几小时桥梁终于撑不住倒塌了。因为桥面太过柔韧，塔科马海峡大桥注定倒塌。只要当初设计师用一根桁架来加强桥跨，其本可以屹立至今，如同全球第一大悬索桥，日本的明石海峡大桥一样。

润扬大桥的长度是塔科马海峡大桥的近 2 倍，但工程师仍旧选择避免用桁架加强的桥面设计。润扬大桥是否会因为"和谐振动"而倒塌？作为当时国内第一大悬索桥，润扬大桥跨距将近 1.5km，全部暴露在强风甚至台风中。但造桥者根据空气动力学的研究结果来设计稳定桥梁的方案。组建桥面板的钢箱梁只有 3m 高，边缘是倾斜的，就像飞机的机翼，让风绕着边缘流过去而不会使桥面板产生晃动。在巨大的风洞试验里，造桥者按照台风的风力对这个设计模型施加强风作用，桥面上下起伏，但没有发生毁灭性的和谐振动。

再过 5 个月就是季风季节，润扬大桥又将面临关键阶段，悬索桥在尚未施工完成的桥面情况下是最经不起打击的。

Mark Ketchum：桥面板的组件由于是吊在缆索上，如果风吹得太厉害，相互之间就会撞击。

建造润扬大桥的人还有一项艰巨的挑战，他们定做了特殊的龙门架来加快施工进度，龙门架沿着桥梁主缆前进，把每一段桥面吊装定位。吊装开始的 90 天后，施工人员终于把最后一块桥面板安装就位。1490m 的跨距，作为当时全球第三大悬索桥的润扬大桥终于完工。随之启动的是艰苦的检修工作。和桥梁的缆索一样，打造桥跨的 2 万 5 千吨钢材也会生锈。但通过一台桥梁内部的小型电动车，检查的工作变得很轻松。这个小车可以让工作人员 18 分钟便能走完全程。桥梁的高科技控制中心就在附近，感应器捕捉了桥梁所有的位移。正常情况下，润扬大桥会持续晃动，因为这种柔性是悬索桥的特征，能够让巨大的结构顺应大自然的意志，而不会发生断裂。

2005 年 4 月 30 日，当时全国最长的悬索桥通车了，进度提前了好几个月。历时 4 年半，共花费了 7 亿美元。这个地区终于有了连接南北的经济命脉。中国准备在 21 世纪跨越长江。在未来的 20 年里，中国打算在长江上再建造 50 座桥梁，中国最壮观的超大桥梁可能还没出现呢。

3.9 全球最长跨海大桥

环球时报：港珠澳大桥主桥全线贯通。连接广东珠海、香港和澳门的全球最长跨海大桥——港珠澳大桥主桥建设工程于 2016 年 10 月 9 日全线贯通。

庆祝这座全长 55km 跨海大桥工程的完工庆祝仪式在珠海举办，也标志着路面和相关工作的开展。港珠澳大桥管理局局长朱永灵说："这意味着大桥主体工程建设进入收官阶段。这座大桥将使从香港到珠海和澳门的通行时间从目前的陆路三小时或水路一小时缩减到半小时的车程。"

港珠澳大桥于 2009 年 12 月在珠海动工。这座 Y 形大桥始于香港大屿山岛，分支连接珠海和澳门。6.7km 的海底隧道和 22.9km 的大桥使用了超过 400000 吨钢材，足以建造 60 座埃菲尔铁塔。英国报纸《卫报》称这座巨型建筑为"现代世界七大奇迹之一"。这个工程施工之前，世界上最长的跨海大桥是位于山东省东部，横跨胶州湾湾口、全长 36.48km 的胶州湾跨海大桥。

港珠澳大桥管理局总工程师苏全科说，该工程已在许多方面都创造了历史。港珠澳大桥位于香港和澳门的机场附近，每天有 4000 多艘通航船舶经过、世界上最繁忙的航道之一，这项工程并没有对邻近船舶和飞机的日常运营产生影响。同时，因为施工区域与中华白海豚自然保护区相重叠，中华白海豚是享受国家顶级保护的濒危物种，其海洋保护也尤为重要。

据苏说，这是中国第一次在开阔水域建立一个沉管隧道。建设过程中第一次使用由两个钢圆筒建立的人工岛来连接桥梁。苏说："该工程特殊的位置和超长的长度都是挑战。拥有一系列世界先进的抗腐蚀和抗震措施，港珠澳大桥的使用寿命达120年。"

主体部分，包括桥面、桥墩、桥梁和钢管，都在工厂生产，然后用大型浮吊在水上组装。海豚形的桥塔重达 2786 吨："两个负载能力分别为 3200 吨和 2300 吨的浮式起重机一起将海豚塔吊起，进行空中转体，创下了一个世界纪录。"

"港珠澳大桥创造的'中国工艺'和'中国标准'将会影响世界市场，"总工程师苏全科说，"如果不是因为'中国制造'的进步，我们不可能在如此短的时间内完工，15 年前浮式起重机的最大负载能力仅为 100 吨。"

Unit 4　德国高速公路

4.3　德国高速公路的设计和维护

Tomas Linder 博士是南巴伐利亚高速公路管理局的主任。

Tomas Linder：德国的公路网是负责联系国内各大城市和重要地区的重要公路，工程水准极高。

德国高速公路能傲视全球有几个原因。首先，这些路比全球大多数公路都厚。根据天气和当地的地面状况不同，公路厚度介于 55 到 85 厘米之间。就算一架 747 飞机降落在上面，路面也不会沉降 1cm，因此它具有坚固的基础，足以应付夜以继日的汽车车流的碾压。这条道路还有另外一项过人之处，就是它平缓的坡度，足以让你能够安全地高速行驶。

Tomas Linder：我们必须遵守很多设计规则，如坡度、曲度和最小半径等，这样才能让车辆高速行驶。

工程人员之所以选择缓坡取代陡峭的山坡来提升行车的速度，是因为路面越接近水平，车速就能够越快。德国高速公路的最大坡度为 4%。坡度对驾驶的影响是：当你在 20% 的斜坡上行驶时，你会在 1km 内爬升 200m；而在 4% 的斜坡上，你要开上 5 倍的距离才能爬升 200m。保持 4% 的坡度看起来似乎很容易，但是考虑德国起伏的地形，从波罗的海延伸到阿尔卑斯山脉，你就会发现这一点也不容易。

德国高速公路的建造者真的要"移山填海"。德国高速公路能称霸全球，最后一个原因就是它有完美的电子系统，这条绵延不断的公路上铺满了侦测器、摄影机、电子路标和通信点，以及电脑不断地监视着各种状况。有些侦测器是硬连线的，有些则是太阳能无线侦测器。它们会反馈任何状况，包括温度、天气、车流量和意外等资料。你在这条公路上的一举一动都会受到严密监视。当然，这项德国工程自然不便宜，为了让所有驾驶都能快乐地进行，德国每年要花超过46万欧元作为每公里道路的维修费，这是美国洲际公路每公里开支的两倍以上。

Tomas Linder： 速度是要花钱的，我们的道路系统非常优秀，它有很好的维护，也有很多科技设备，驾驶员都能快乐地开车。

4.4　完美监控系统

事实上，德国高速公路偶尔也会有超速陷阱。在12000km的道路中约三分之二没有速限，这对飙车族而言已经很不得了。然而，德国交通警察在这条高速公路上会以快制快，他们的车辆通常比被追逐的车辆更酷。交警配置的汽车是外国汽车收藏家的梦幻车队，奔驰、宝马、奥迪，这些车的马力都足以和赛车媲美。这些车有统一的警车标配，如无线电、电脑、旋转轴和摄影机等，不过它的引擎盖下还另有玄机。除了有额外配备设置的加强避震器，这些车也是少数不必装250公里限速器的车辆。德国警车需要额外马力的原因之一是普费芬豪森有个制造高速车的家族。很多驾驶人都认为最适合在高速公路行驶的就是鲁夫家族的保时捷。

鲁夫家族从1939年就投入汽车业，从保时捷的维修公司发展成改装公司，1981年阿洛伊斯和东尼亚，开始以保时捷自行生产汽车，他们的手工车被认为是世界上最高级的车种。

阿洛伊斯和东尼亚： 德国是唯一拥有无限速道路的国家，我们则专门生产举世无双的汽车。德国车一直是世上最佳的汽车。我们鲁夫家族会采用最佳的汽车，也就是保时捷跑车，把它改装得青出于蓝而胜于蓝。

R涡轮是鲁夫家族的最高配置汽车，它的底盘来自保时捷996涡轮，其他部分则是经过重新设计，它是德国高速公路迷的梦幻车。四轮驱动的R涡轮，能在4秒内从静止加速到时速100km，极速时可达时速350km。这到底有多快呢？大型喷射客机的起飞速度约为290km/h。基于此，德国高速公路的巡警该如何对付这样的快车呢？他们靠的是巡警沿用已久的工具。

首先是技巧，高速公路警察会拦在目标车辆的前方，这样就能让例行的拦车不至于演变成高速追逐；接着就是科技，大多数运营车辆都配备速度记录器，警车查看车辆的速度记录后就能开超速单。但有时，传统方法并不适用于这条超级道路，必须以非常手段让大家减速慢行。德国交通警察如何进行取缔超速的训练？秘密在于他们会玩"电动游戏"。这"电动游戏"是警车专用的驾驶模拟器，位于稣兹巴克的警察训练中心，距离慕尼黑北方约200km，这种高速驾驶模拟器可能是绝无仅有的，它就像飞机飞行模拟器一样，只是它模拟的是警用宝马。这个模拟器有个巨大的全景屏幕，为求逼真的效果，它的后视镜也有影像。模拟器能模拟出各种驾驶状况，从小镇、乡间道路，到城市塞车和高速公路上的高速奔驰。受训者最需要知道的就是人们的反应。在超高速追逐中，警笛并没有什么用，因为驾驶人在全速前进时听不到警笛声，当驾驶人从后视镜看到蓝色闪光时，警察已经要超越他们了。警察接受充分的模拟器训练后，便能在几秒内作出反应，并训练出远优于一般驾驶人的驾驶技巧。

警察的第二道防线，就是机车警察和他的超级机车。一种机车类型是宝马 K1200GT，极速可达 240km/h，它能在 3.7 秒内从静止加速到 240km/h。骑机车在德国高速公路巡逻，是一项危险的工作。这条公路的大多数驾驶员都有安全带、气囊和滚动底盘作为保护，但机车警察则只能靠安全帽、皮衣裤和自身驾车技巧。有时在这条超级道路上行驶，上述安全保证仍不够，因此警察必须有另一项更隐秘的工具，即高科技监视装备能让驾驶人遵守交通规则，就算警察不在现场。这种方法叫照相执法，但称它为公路监视者也许更合适。停在路边的伪装车辆和路标以及地下道，都装置了摄影机。这些隐藏摄影机会拍下超速的车辆。影像经过处理后，底片会被扫描进电脑系统，以数位处理方式增强影像的清晰度。接着它会扫描违规车辆的驾驶员脸孔，并加强影像。最终两张照片拼在一起，连同罚单寄给自以为没被发现超速的驾驶员。

照相执法能奏效是因为它很隐秘。我们都知道超速司机看到警车自然会减速。但警察知道驾驶者常会心存侥幸，有时他们一超越警车就会开始加速。在这种情况下，就要靠便衣巡警了。这一群便衣警察，用的是高科技装备，例如隐藏式摄影机和敷光测速器。这些便衣警察在车队中穿梭，在高速公路上寻找违法者。一旦发现超速者，他们就会把蓝灯固定在车顶，去尝试拦下违规者。这些警察确定对方看到他们后，就会挥牌要求停车，让驾驶员跟着他们停下，接着便会开出罚单。

Wolfgang Steiner：我追过最快的车，是在限速 120 公里路段开到时速 245 公里，这样的超速会受到 375 欧元的罚款。违规驾驶人也会吊扣驾照三个月。

各地便衣警察的工作都很"刺激"，在德国高速公路这种刺激有时会变得很惊心动魄，甚至很危险。

Christian Renftle：我们在高速公路上行驶，也是十分危险的。其他驾驶人并不知道我们是警察。他们有时会拦在前方让我们减速，这真的很危险。

虽然德国高速公路是世上最快速的地方之一，但在德国警察的严密监控下，它仍是世上最安全的道路之一。

4.5 高效道路救援组

直升机救援飞行员 Roland Benning 正在进行例行保养，让直升机加满油准备进行下一次任务。

Roland Benning：我们此刻看起来很轻松，但随时都可能出现 180 度的变化。

此时，这名飞行员就遇到了他所说的"180 度"变化：因为有辆超速车失控了。Benning 是德国汽车联盟的飞行员，这个汽车联盟又被称为 ADAC，是德国高速公路的生命线。ADAC 是个强有力的汽车协会，负责从补胎到空中救援等各种任务。这个道路救援组的服务范围远远超出了他们的职责，因此被誉为"黄色天使"（因为救援车的车身颜色是黄色的）。ADAC 旗下有 1700 个运行维护站，并在德国各地有 36 个直升机救援基地。Martin Hauser 是位于慕尼黑旁兰兹柏 ADAC 道路维修中心的组长。

Martin Hauser：如果车辆在高速公路上就抛锚了，他们就可以致电维修中心，我们接到电话后会问一些问题，如地点、车型、故障情况等，还有是否需要拖吊或直接能在现场维修。我们将资料输入电脑后，调度员就会收到资料，并就近派出人员。

ADAC 的救援服务可不只是拖吊而已，"黄色天使"会在现场修车，八成的车辆能在现场修复好。这些救援汽车里就有个维修站，他们会在必要的时间展开修理绝活，例如在路肩钻孔和焊接等。"黄色天使"的每辆车都是定做的。Alois Stiegler 是 ADAC 的主要定做商，这位天才改装车上修理站已有 25 年了。

Alois Stiegler：我们每年约要把 200 到 250 辆原装车改装成运行维护站，每辆车大约花 55 小时改装。

这些改装完成的"黄色天使"车，满载着各式各样的设备，从简单的扳手到先进的电子检测工具。这些设备修复了德国高速公路 83% 的抛锚车辆。

4.7 高速公路控制中心

回顾历史，我们发现德国高速公路有着高速公路的"血统"，它的前身是条赛车道。那是一条从 1913 年开始建造的实验车道。这条车道长 19km 位于德国南方，起初被当做赛车跑道

Cornelia Vagt-Beck：德国高速公路系统，从 1930 年后开始大幅扩张，主要是因为人们发现在平坦的高速公路上开起车来轻松多了，高速公路可以让车速更快。

截止 1942 年，高速公路已经修建了 3900 多公里，并连接起德国大部分城市。在第二次世界大战结束后 8 年，德国高速公路又恢复建造，这条巨兽般的公路从此不断增长。但这条路可不只是绵延不断的厚水泥路而已，这是一条"会思考"的路，它尖端的基本设施就像它的车速一样令人尖叫。

德国高速公路其中一个控制中心位于慕尼黑。它负责监控长达 1200 多公里、德国高速公路最忙碌的路段。控制中心的主要工作之一就是预防"大塞车"。

巴伐利亚哥公路的每天车流量为 45000 辆，慕尼黑每天的车流量为 140000 辆，节假日时更高达 180000 辆。这些中心监控着拥挤的交通，无数摄影机监控每公里的高速公路。在目视监控鞭长莫及的地方，则有侦测器侦测路面状况。这些高科技可以察觉声音、动作和热度，大概是世上最聪明的高速公路了。这些资料会被传到控制室电脑，电脑会处理并传送相关信息。由于侦测器不间断地观测高速公路，可以想象将有多大的数据流汇入控制中心的主机中。侦测器控制交通塞车的原理如下：侦测器在慕尼黑外发现塞车的微兆，也许是摄影机发现车速变慢了，也许是雷达测速器发现车辆几乎没有前进。控制中心的某部电脑收到资料后，便会对缓慢路段的电子路标传送指令。因此，电子路标的符号会突然改变，路肩成了新增的车道，车辆又开始以高速行进。所有这一切都是电脑控制的。除了改变车道，控制中心也会采用开放路肩来简单有效地疏散车流。在交通尖峰时段，通过远程操控路肩会转换成新车道，以疏散拥挤的车流。除非是因为意外、施工或者警方授意，这套安全的系统不需要人力介入。

Tomas Linder：警方需要改变路标时便会通知我们，他们会致电控制中心，控制员接着就会按照警察的指示改变路标。

4.8　精准设计和精心维护

控制中心是在幕后指挥高速公路的"大脑",维修工作则每天都会处理具体的事物,让这条道路保持光鲜亮丽。维修是德国高速公路备受称赞的主因。每天都会有无数专用车辆在德国各地穿梭,检查道路是否安全,道路状况是否最佳。

Tomas Linder：他们会在冬天除雪,在夏天除草,修理路标和安全栅栏。一旦有车祸发生,维修人员就会赶到现场协助维修。

Klaus Seuferling 是慕尼黑维修中心的主管,他对自己的任务非常了解。

Klaus Seuferling：我们的第一要务就是确保往来车辆的安全,并保持交通顺畅。

道路维修的主要工作就是处理大批高速行驶的车辆,维修工人启用特殊的车辆保障道路的通行安全。

Klaus Seuferling：使用那种设备的工作人员,处境有时会非常危险。而驾驶人也必须提高警觉。因此,为了保护维修工作人员,必须依靠特殊的车辆。这种特殊的车只有一种用途——防止其他车辆撞上工作人员。它能让车避开施工中的车道,车上还配备了特殊的吸震物质,这些物质能吸收高速碰撞的力,让工作人员不至于受到重伤。

维修人员还有很多其他专用车辆,例如一辆乌尼莫克,采用奔驰的车头和各式各样的连接配件,能胜任很多工作。这辆乌尼莫克的专长是清洗路旁的标志。尽管德国高速公路配备有大批各司其职的昂贵机器,但其中最佳的维修"利器"还是巡视人员。

Klaus Seuferling：德国高速公路的每位工作人员的任务,就是检查负责路段的状况。他们每天都会开车在德国高速公路上穿梭,看看哪里有问题和哪里可能发生什么,并寻找精准的修理地点。

德国高速公路适合行车的最后一项因素就是它的建造方式。这条道路建造得很坚固,大多数路段都是刚完工不久,并不需要太多维护,因为它几乎不会损坏,不会出现裂缝和坑洞。这条路几乎没有坑洞的原因之一,就是因为它非常厚。德国高速公路的平均厚度为 70 厘米,对比美国高速公路的平均厚度约只有它的一半。水泥会热胀冷缩,路面愈薄就愈容易受环境影响而碎裂。

德国高速公路不会裂开的另一个原因,就是采用易排水的结构。让人意外的是:水竟然是破坏高速公路的主因。雨水对道路造成的破坏,要比无数速度高达时速 200 公里以上的轮胎碾压还大。工程师以倾斜路面防止雨水侵蚀,以下就是它的原理:雨水会沿着坡度为 2.5% 的路面流到路的一边,再渗过一层透水的水泥,流进水管网路,最后流进储水塘。雨水被储存在这里,免得其中的污染物破坏环境。

尖端的科技,创新的工程学和一丝不苟的维护,让德国高速公路充满活力。你不是在这条路上开车,而是和它配合着开车。

一名汽车司机：这是一条精心设计的公路,这条道路的品质很棒。你常会看到维修和养护工程。这条路维护得很好,开起车来很舒服,它真是精心设计的公路。

Unit 5　京沪高速铁路

5.2　车型设计大赛

这项创纪录工程的总设计师负责设计列车。设计团队在其中倾注了很多心血，他们希望最终的设计能够在技术上有杰出的表现，同时也能代表中国文化。

日本以其子弹列车首度突破 200km 时速障碍，最新的设计是鸭嘴兽外形。法国高速列车于 1979 年突破 300 公里时速障碍，它拥有扁平低斜的车头。目前中国最快的列车 CRH3 时速是 350 公里，但那是中德技术合作的成果。相比之下新的 CRH380 列车为纯中国制造的，它需要特殊的外形设计才能匹配这个速度。

设计团队在想世界上有哪些东西具有这么快的速度，比如说马的速度很快，高山瀑布，以及发射的火箭。他们放开想象，想出了 20 种独特的设计。以空气动力学作为评判标准，来评判列车头的设计，同时考虑设计方案是否能表达出中国文化的精髓，5 种方案最终脱颖而出。

经过第二轮的比选，最终方案确定了下来。这个胜出的模型即使在静止时也看起来具有速度感，它的外形非常流畅具有动感，空气阻力和噪声都控制得很好，驾驶室和前部细长的比例都非常完美。车头方案选定后，该着手建造了。

最具困难的挑战之一，就是要让列车足够坚固以保证匹配新的速度。在时速 380km 下，列车前方的空气极度压缩，就像车头在不断地撞穿砖墙。如果车头不够坚固，光是空气压力就能让它支离破碎。加大车壳厚度并非上策，因为如果列车太重的话就无法达到预定的时速。最终，解决的对策是采用空心车壳，兼顾坚固和轻盈。应用在空心车壳的焊接支架能增加强度的同时，不增加车壳厚度。

时速 380 公里比大型喷气式客机的起飞速度还快，设计团队需要确保火车平稳地行驶于轨道上，设计师在车头两侧加上凸起的肩线。这些肩线像翅膀一样，但方向却相反，为了让气压将列车往下压而不是往上升。这些特征和车头其他部分一样都是平滑曲面，可把空气对列车的阻力降到最低。当然空气并非列车面临的唯一问题，列车上最脆弱的部分是挡风玻璃，必须能够应付任何状况。

5.3　全球最强劲的玻璃窗

在中国国家安全玻璃检验中心里，设计师正在寻找世界最强韧的玻璃，可以用在世界最快的列车上。通常试验仅在时速 350km 的条件下进行。因此设计者们担心现有的玻璃无法承受 380km/h 时速下的气压，尤其担心飞鸟。当高速列车开来时，由于速度太快，鸟根本没有时间逃离。而当它们有反应的时候，已经撞上了车窗。

第一片待测玻璃固定于坚固的夹具上。技术员准备"射弹"，即采用死锥模拟飞鸟。一座 9m 长的大炮发射死锥，发射速度达到 380km/h 以上。撞击的结果非常可怕（玻璃瞬间被击碎）。他们必须找到足够坚韧的玻璃，否则整个工程注定失败。

第二片待测玻璃不是普通玻璃，而是一片超薄玻璃夹层包裹着树脂内层，专为吸收高速撞击的动力而设计。这次测试非常重要，如果新玻璃能承受撞击，便可直接用于时速 380km 的列车设计中。"炮弹"发射后，玻璃似乎没有损坏，经过详细检查发现连一条裂缝都没有，它完美地匹配了新列车的时速要求，意味着设计师可以更进一步，为创造世界最快列车的梦想而努力。

5.4 转向架保证列车完美平衡

在当时，最快的轨道列车极速为 350 公里，听起来将速度提升到 380km/h 或许并没有增加多少，但速度影响一切，设计师必须重新设计许多零部件。

受高速影响最大的列车零件之一，就是轮组或者转向架。最适合的转向架设计过程包含无数扭扯和测试设计师非常辛苦。前后共花了 8 年时间不断修正设计。花费这么长的时间主要是为解决一种奇怪的现象——能动。

在高速下，如果转向架没有完美平衡，整列列车可能会左右剧烈摇动，这种现象称之为能动。能动的发生是因为每个轮子表面不是平的。他们有点倾斜，列车会在斜度间做钟摆摇动。但这种斜度不能被铺平。因为斜度让列车能够转弯。其原理如下：列车进入弯道时重力倾向一侧，把外侧车轮推向斜度顶点，而内侧车轮则处于斜度底部。由于车轮转动直径不同，而将列车导入曲线路径。

不幸的是在直线轨道上，这些不平衡就会导致车轮左右摆动，速度越快，摆动就越严重。

经过 8 年的研究，解决不平衡的新型转向架进入测试阶段，它将被置于 420km/h 的测试环境下。如果没有明显的振动，试验就认为成功。

测试机台已经全速运转，测试机台上的车厢必须稳若泰山才算通过测试，出现任何微小的移动都算失败。最终，新型转向架顺利通过测试，可以应用于新轨道上，能动的问题终于得以解决。

5.5 支撑全球最快列车的铁路桥

在中国各地的试验室持续对新列车的各方面进行测试时，负责建造铁路的团队碰上了难题，他们必须跨越亚洲最大的河流。为了解决这个问题，他们需要建造一座足够坚固的铁路桥。

第一条京沪铁路建于 1908 年，一百多年来一直是两座城市间的交通主干。但这种情形即将改变。它将被世界最快的铁路取代。但要完成这条铁路，首先要跨越世上最繁忙的河流。这是整个设计中最艰巨的工程挑战。

浩瀚的长江将中国分为南北两部分，铁路跨越点的江面宽度达 1.6km。上海和长江三角洲工业区间每天约有 3000 艘船通过。文武松负责设计这座大桥。

文武松：这是我所建的最困难的一座大桥。水越深，整个任务就越困难。

决定这座桥需要承载 6 条铁路线路，这让文武松的工作变得更加困难。6 条线路包括：北京-上海线，上海-武汉线，及附近城市南京的一条地铁线。文武松的首要任务是决定采用何种桥梁设计。

文武松：因为这里水深大约 51m，河上航运非常繁忙，这种情况通常会采用宽跨，就像那个公路桥一样。

宽跨桥梁是理想的选择，因为它不需要建造许多水下支撑。虽然这种桥型很适合公路交通，但不适合又快又重的列车。最符合这项工程的设计将是连续桁架桥，但它依赖混凝土桥墩作为支撑。而在足以淹没 17 层楼的深水里，建造这些桥墩对于工程师来说将是艰巨的挑战。

文武松： 还没有人尝试过在 51m 的水深建造这样的桥梁。

文武松的对策是采用叫做围堰的沉入式平台，让工人能更接近河床，就像一个巨型的水桶一样。宽 80m，重达 6000 吨，由坚固的钢材制成的 "水桶"，利用全球定位系统进行定位，其误差不超过 5cm。它们提供了干燥的沉入式工作平台，不受江面交通的干扰而且更接近江底。工程师在每个围堰底部挖洞以容纳钢套筒，套筒往下延伸至江底的淤泥层，再开挖深孔，嵌入岩床，灌入混凝土作为桥墩，最后将围堰本身灌注混凝土，这样桥梁就能坐落在宽阔的平台上。

围堰的面积等同于 7 个篮球场，能承受 80000 吨压力，加上 6 列列车同时通过。完成平台的工作后，就能开始建造桥梁主结构，利用爬升式起重机将构件逐渐吊至定位，再用螺栓拧紧。经过 40 个月辛劳的工作，荣耀的时刻终于要来临，工人要将最后的螺栓拧紧。但此时天气变化却使计划无法顺利进行，白天温度过高使桥梁过度膨胀，导致螺栓孔无法对齐，工人必须等到日落时温度下降使桥梁收缩。

文武松： 我们看着孔缓慢地移动、对齐。当所有的桥孔全部吻合时，现场所有的人都很兴奋。

完成的铁路桥是人工奇迹。在当时，它是世上轨道最多、跨距最大且负荷最重的铁路桥。但它不是最长的桥梁，这项荣誉属于京沪铁路的另一段。

为了保护农田，工程师们神奇地将百分之八十的路线以桥梁架高，这也能避免不速之客，例如牛、羊等牲畜误上铁路。从丹阳到昆山之间是最长的一段陆桥，延伸长达 164km，是马拉松距离的 4 倍，这样就使它成为当时世上最长的桥梁，长度是第二名的两倍。

5.9　防止列车共振现象的损害

在这专门为北京-上海高铁设计的列车正式上轨道之前，还必须通过一项重要测试避免共振现象的发生，这种现象对任何建筑来说都会引起毁灭性的灾难。共振是一种非常剧烈的振动形态。设计团队必须采取措施对其进行防范。

共振是物体在特定频率下最严重的振动倾向，例如，某根音叉在 440 赫兹时将产生共振。共振的危险性表现在物体在其自振频率时的反应状态。

1940 年，美国刚运营的塔科马海峡大桥受到灾难性的破坏。吹过的风和桥的共振频率正好吻合。风的能量直接助长桥梁的振动，使它振动过度而坍塌。为了避免类似这样的灾难，列车的各个组件被设计成不同的共振频率，并通过谨慎选择材料和连接的方式来避免共振。这样的设计对车头和转向架尤其重要。

此时正在进行的是车头的试验。列车头的各个角落都放置振动锤，一名技术员慢慢转动刻度盘，改变它们的振动频率。当振动锤的频率和车头共振点恰好吻合时，它会剧烈振动。这种振动肉眼很难看出来，因此在车头上放置一瓶水来观察。为了避开转向架的共振频率，车头的共振频率必须高于 6 赫兹，但也不能太高否则会干扰其他零件。试验的目标是 17 赫兹。试验从 13 赫兹开始往上加，设计团队仔细观察是否有共振的迹象。

当达到 17 赫兹时，列车开始剧烈振动。每个表面都剧烈颤动，整个车头发出类似蜂鸣的声音，这就是车头准确的共振点。

Unit 6　英法海底隧道

6.2　主隧道和附属隧道

实际上共有三条隧道——两条主隧道和一条维护隧道。北边的隧道供从英格兰到法国的列车行驶，南边的隧道则反方向行驶。较小的维护隧道则是位于两条主隧道之间，是整个隧道非常关键的设计，因为它既是事故发生时的疏散隧道，同时也是实施维修工作的主要通道。汽车也可以通过维护隧道从英格兰开往法国，但基本上没有普通汽车试过。

Richard Dance：检修车辆上有导向系统引导车辆行驶，所以你有时甚至可以放手驾驶。路上埋了两条电缆控制方向，检修车辆沿着电缆行驶。你只需操控脚踏板实现刹车或前进，其他都是自动化的。现在我们正位于无灯地段，所以我们能够加速到50km/h。

每隔382码就会有连接维护隧道和主隧道的横向通道。另外主隧道还与一系列被称为活塞形管路的通道连接，让两条主隧道之间的空气流通。维护隧道不对外开放，事实上它也不是完全敞开的。维护隧道里的空气压强高，造成维护隧道高压密闭现象，最主要的原因是维护隧道是乘客遇到事故时的主要逃生通道。让它保持在比主隧道压力高的状态，能够确保发生事故时，烟雾和污染的空气无法进入维护隧道。同其他工作者一样，Richard对地底下完全不同的环境记忆犹新。

6.3　隧道的挖凿工程

Richard Dance：地下环境非常肮脏，但这是一个完全不同的工作环境。所以你必须确保自己的安全。在六架钻孔机同时钻孔的情况下，环境非常险恶。其中三架钻孔机往法国，另三架转向内陆。

承包商决定从两头同时开工：一端从莎士比亚悬崖多佛开始，另一端则是法国海岸的桑加特。六架钻孔机从莎士比亚悬崖同时开工，三架驶往内陆福克斯通，另三架穿越海底，前往法国。在法国那端也采取同样的挖凿工程。但有些路段是靠工人手挖的。

Richard Dance：看这些由圆形隧道转为方形的部分，这里是吊出钻孔机的地方，是封口地带，因为它是从地面上挖凿的，土壤由此被运到隧道顶部。随着我们前进，截面形状又发生了变化。我们又回到了圆形隧道，这一部分是人工挖掘的，没有使用钻孔机。在这，两条主隧道之间出现了维护隧道，它慢慢下降并沿主隧道方向延伸，接着又上升，并在你离开之处结束于主隧道的左侧。

在法国悬崖上没有方便挖掘的着手点，所以在开凿之前，他们不得不在桑加特挖掘入口通道，钻孔机才得以从洞中降运下来。隧道漏水的时候，这是一项艰巨的工作。人们必须在冰海水中连续奋战好几天，甚至好几个星期。每一架隧道钻孔机都是一个小型工厂，旋转头切割白垩，水力活塞推动机器前进，在凿头后面，紧跟着把预备好的隧道内壁放置到合适位置，废弃物由火车运回海岸。

他们在海底两个位置挖掘了巨大的洞窟，以便火车能够在两条主隧道之间穿行，这些称作转线洞窟的通道是隧道建造者最伟大的杰作。英格兰那端的转线洞窟在距离海岸七里的地方。维护隧道的钻孔机走在主隧道钻孔机的前面，当它到达海底下方60码的转线洞窟时，便开始减速向外挖掘。在那里，隧道挖掘者挖掘了一系列通道组成转线洞窟，也是从那里开始进行人工挖掘：他们首先挖出顶部，然后是两边，最后形成两条隧道。在大型钻孔机到达前，地下已经有长178码，深16码的洞穴。挖掘在创纪录的速度下进行，隧道工作者可以腾出时间，看到主隧道钻孔机突破，进入转线洞窟冲往法国。

对于承包商来说，转线洞窟是唯一能够看到整台钻孔机运转情况的地方。为了保证进度，机器的最快运转记录是每小时300码。转线洞窟看上去像两条隧道，是因为其中有一堵可拆卸的墙把南面和北面分开。

当墙壁撤掉时，北隧道的列车就能够行驶到南隧道。而在法国段，可以反向穿越。这样，隧道就可以分成六个部分，每一部分都能够单独维护。为了维护洞窟墙壁，工程师在白垩上钉了好多很深的钉子。他们稍稍移动顶部以实现密切地监视。当墙壁松懈，他们就会涂上防水内壁，喷上混凝土进行加固。

同时，隧道钻孔机（TBMS）继续向法国前进，隧道钻孔机使用最先进的激光技术操控，机器在30里的路程中，从来没有偏离目标一英寸以上。1990年12月维护隧道的竣工载入了史册。为了与时间赛跑，也和不断上升的银行利息角逐，挖掘夜以继日地进行着。随着挖掘的深入，隧道里的人、装备以及废土运输到内陆的距离也就越来越远了。

1991年6月28日，主隧道的最后连通标志着第一阶段建造的完成，比1985年预计的竣工时间提前了两天。在短短的三年时间里，承包商挖掘了100里的隧道，取得这一佳绩的自豪感是无法用言语形容的。他们最快的挖掘速度达到每月一里以上，这是前无古人、后无来者的奇迹。

总共有11架钻孔机参与了海峡隧道的挖掘工作，当挖掘完成后，它们各自面临着不同的命运：两台被出售了，有几台已损毁，还有一台沉入了茫茫大海，另外一些被作为纪念品陈列在欧洲隧道公司的办公室。还有一台被遗弃在福克斯通高速公路旁，提醒着人们挖掘洞穴工作只是完成了所有工程的一半。

6.4 运输体系

真正的挑战是把洞穴转变成有效的运输体系，它不能像任何老式的运输一样，它必须是最棒的运输方式。在当时，隧道公司已经负债累累。如果想偿还债务，修建的隧道运输体系必须是一流的。

相对来说，将隧道与英法现有的铁路系统连接会比较容易些。从伦敦到巴黎或布鲁塞尔乘火车只需三个小时。但比较困难的是运输从福克斯站进出的汽车和卡车。不仅隧道本身受到了大众的瞩目和很高的评价，隧道两端的汽车终点站也代表了运输设计和技术的重大进步。英国端终点车站，有运输客车所需的串联设计，设计师从安全保障到高速公路的入口，都进行了详细的设计，而其中最重要的问题是速度。

隧道运输的关键点是保证小汽车、卡车和公共汽车都能够顺利地出入穿梭列车。如果穿梭列车想要跟海峡渡轮竞争，全部的路途包括装卸时间必须控制在一个小时以内。这意味着要有两样东西——即特别制造的运输工具和特殊的月台系统。没有这些隧道将无法有力竞争。

福克斯通终点站上方的主控室里有交通管理员，负责确保汽车以最快的速度进出穿梭列车。听起来这工作很简单，只需在最短时间内把汽车排好并运上列车。绝大多数情况下，92%的英法隧道穿梭列车能够准时。但是人们对这一系统还存在争议。在穿梭过程中，乘客们必须坐在车上，他们看不到周围的风景，也没什么可做的。如果发生意外也没有有效的撤离方法。举个例子，如果发生了火灾，乘客首先必须离开他们的汽车，走下穿梭列车内部的楼梯，进入隧道后还要在黑暗中寻找出口。如果隧道烟雾弥漫，将使逃生难上加难。

不同的列车其行驶速度也不同：客运列车时速100里，货运列车时速60里，客运穿梭列车时速在90里。在前序列车和后序列车之间，要保持3到4分钟的行驶距离。无论任何情况，一条隧道都不允许有超过五辆列车在同时行驶。月台到月台之间大约有35分钟的路程。

6.6 排水设施

从英格兰到巴黎的铁路线上，一列旧的火车正在缓慢通过，这些线路已有超过150年的历史了。只有进入隧道后列车才能真正提高速度，当到达隧道口时，这感觉好像在下坡一样。承建者在设计隧道的时候决定从白垩内部，即白垩核穿越。这是钻孔最容易且最稳定的区域，使得隧道蜿蜒穿越海峡而延伸。隧道内有三个低点，作为排水区域。令人惊讶的是，隧道的漏水居然是故意设计的，因为不渗漏隧道周围岩石中的水压就会大到无法承受。英国抽水站位于海底下方45码处，它的上方还有60码的海水，岩石和隧道内壁都是多孔渗水的，水从隧道渗出并沿着隧道内壁背部下流，汇聚在铁路下方专门建造的集水沟中，再靠地心引力流向抽水站，每天都要抽取9万加仑以上的海水。

无需给主隧道注入空气，因为进入隧道的列车本身会带入空气。任何移动的物体，尤其是列车，都会带动其后的空气流动，在隧道中这可能引起气压增加。为了避免这种现象，并使两条主隧道之间空气流通，承建者建造了第二组的连结通道。它不像横向通道，这些"活塞形"管路并不是封闭的，空气能够在其中自由流通。

但维护隧道的情况就不同了，为了安全起见它是全密闭的，并能把气压控制在比主隧道高的状态之下。在法国的桑加特，原先用来升降隧道钻孔机的通道，现在另作他用。这些大风扇吸入空气，空气被推入这些管中，经由露天的输送管，到达通道并进入增压的维护隧道。每个隧道的确切体积可以被测出，维护隧道一共长30英里，直径16尺，为了确保其气压比主隧道的高一个大气压，每秒钟必须注入220立方码的空气。类似的装置也设置在英国的莎士比亚悬崖底部。

6.8 消防保护措施

隧道基本上都保持在黑暗状态下，只有当人在某段隧道作业时才会打开隧道灯。一辆消防车经过，维修经理Richard继续向前开，前往出口。

Richard Dance：我们现在马上就要进入英国入口的气闸，为确保维护隧道的压强稍高于主隧道，它设有双重闸门，保证在主隧道发生事故的时候，维护隧道能够成为安全避难所。我们需要获取进入气闸的通行许可。门打开，我们可以进入了。我们身后的气闸关上以后，前方的出口才会打开，我们得以重新呼吸新鲜空气。

当然打开这道门也需要获得通行许可。

Richard Dance（通过对讲机获取许可）：这里是 63 号，我们现在正在英国气闸，请求离开维护隧道。

使用密封气闸系统的操作方式是缓慢沉重的，但理查从未忘记安全的重要性。

Richard Dance：在安全的操作状态下，才能够保证你的安全。而如果你碰运气，很显然就容易出事。

建造隧道时共造成 11 人伤亡，法国段 2 人，英国段 9 人，而到目前为止，只有一名工作者死于隧道运行中。但是危险是无处不在的。欧洲隧道公司认为每个人都应对安全问题有足够的警惕性。隧道中有许多人和车，一旦发生火灾是极其危险的。消防队坚持在隧道开展 24 小时巡逻，保证 20 分钟内能到达隧道的任何地方。

但有时候连这样的巡逻也捉襟见肘。当时在 1996 年穿梭列车中的幸存者回忆起那次事故，那也是迄今为止唯一的一次事故。

幸存者：当时车内充满了烟雾和有毒气体。我们能获救真是一个奇迹。如果多给六分钟的时间，我们就能全部被安全救出。

1996 年 11 月发生的惨剧对欧洲隧道公司是一次严重的挫折。火灾发生在一辆沿着法国到英国路线的货运列车上，列车刚驶入海底几里处。货运车不像客车，它是敞开的；一辆卡车着了火，火势和烟雾会迅速扩散，紧接着引起的混乱使安全措施失效。司机无法使列车减速，用来清除烟雾的风扇也没有打开，烟雾迅速迂回到前面的餐车。事故共造成 34 人受伤，8 人由于吸入过多烟雾而被送往医院。幸运的是，没有人员死亡。

幸存者：两分钟内我们到达地面，呼吸到了新鲜空气。

消防队员和货车驾驶员被送到错误的维护隧道入口。400 名消防队员，足足花了 6 个小时才把火势控制住。隧道内的高温甚至将部分列车和铁路焊在了一起。

消防队员：我在这一行做了 20 年，但从未经历过这样的事情。当时我的防火头盔非常坚固，从外观看上去似乎完好无缺。但如果你观察里面，就会发现头盔其实已经（被热得）起泡了。

Unit 7　芝加哥地底城

7.2　阿姆斯特丹运河系统下的地下空间

阿姆斯特丹已经开始规划向地下扩展空间，因为上面完全没有土地了。在荷兰这座城市里，路面电车、行人以及 55 万辆自行车，拥挤在狭窄的空间中，日益增长的汽车数量严重影响了交通。

Bas Obladen：阿姆斯特丹是座美丽的历史古城，而城里的交通很拥挤，你无法尽情享受这座城市，因为到处堵得水泄不通，包括河道堵塞、码头拥堵等。

古代的城市规划者从未料想到这些美丽的蜿蜒小径现如今已经车满为患。现今，有建议将汽车赶出街道呼声，但如果既想减少汽车同时又维持阿姆斯特丹的历史风情，就要依靠它自古流传下来的特色——运河系统。

这项计划就是在全城 65 里长的水道下方建构地下城市。停车场、电影院和商店都建构在数里深的地下，上面刚好全是河水。

Bas Obladen：利用运河空间绝对是最佳选择。因为你把原本一平方米毫无价值的空间，巧妙转变为六平方米炙手可热的空间。

这项计划将改造超过 20 平方英里从未利用过的空间，提供数千辆汽车的停车位。但在阿姆斯特丹建设工程必须考虑当地建筑特色。阿姆斯特丹的街道无法容纳大型建筑机具，有些街道过于狭窄，导致大型卡车根本开不进来。

即使施工人员可以从街边进入工地，建设工程也会由于接近运河岸边的房屋而具有危险性，这些房屋颤颤巍巍地竖立在水边。

"如果说在芝加哥，你挖个洞就能在上面盖房子。但在阿姆斯特丹并不容易，因为这里的土壤很多人称之为软水土。"

阿姆斯特丹的软水土或叫沼泽土使建造大型建筑物几乎不可行，所以这里的建筑很少超过三四层。大多数运河周边的房屋，仅靠插入软水土的木桩支撑，建设过程中引发的振动很容易造成惨剧。远离房屋的只剩运河，其宽度也够容纳工程机械。但考虑到运河里全是水，工程师提出了独特的解决方案。

阿姆斯特丹的运河系统有 16 道闸门可以随时关闭，闸门一关运河里的水就能排干。

"我可以想象通常大家都会认为，把运河的水排光是最困难的。但其实在技术上，这是最简单的一项工作。"

等运河一干，施工人员就可以开工了，往下挖 30 尺后封闭新表层，这个表层将同时作为地底建筑的屋顶和运河底部。这个方式不但大幅减低对运河旁数千栋房屋的伤害，也不会让整座城市交通陷入滞留状态。

Bas Obladen：因为我们在运河区域内部施工，不会对大众造成过多不便。

稍后只要打开闸门就能让运河恢复。

Bas Obladen：我们完工之后，运河看起来就和现在一模一样，只是路面上不再有汽车，你就可以尽情享受整座城市，欣赏这些美丽的古楼。

这项计划将使数千辆汽车从街边消失，这项计划实施成功的话，将鼓励其他城市采取类似的方式。但却面临一个问题：很少有城市是靠运河穿梭市区。在芝加哥根本不可能在开挖路面时，不对全市交通和居民造成干扰。为了芝加哥美好的明天，工程师必须想其他的办法，从城外解决问题。

7.4 挖掘岩层的高效技术

施工人员进入地下 400 尺深处挖掘最新的岩层，他们的目标是在一天之内开挖出数千吨岩石。他们采用的正是所谓的"钻爆法"。这个 385 匹马力的钻头能在岩石中掘出 16 尺孔洞，每次可以掘 59 个洞，由铵油炸药装药机负责将洞里填入火药。装药机的两只槽中装有五千磅的铵油炸药，也就是包覆着燃料油的硝酸铵，这些炸药经由空气软管注入洞内，接着设定循序引爆。

Charles Dowding：这是个缓慢讲求方法的过程。钻爆法必须以严格的顺序进行，包括钻掘，埋药，净空，引爆，通风。难度在于净空时要安排人员撤离，使完成每一个施工流程至少要耗费四到八小时。

安全问题是首要任务，这个做法奏效了。这一回合所使用的 1100 磅铵油炸药，炸开 1400 吨的石材，挖开 14 尺厚的墙面。钻爆法可将原本厚实的岩层改造为一处开放空间，但同时有些结构问题需要立刻解决。

Charles Dowding：开凿大型洞穴会造成应力集中，使得缺乏加固支撑的岩层过度负荷，而且洞穴越大越深，应力集中就越大。

为了补强加固，需要派出另一台钻掘机掘出垂直孔洞，填入环氧树脂和钢筋加强支撑洞顶。爆炸之后，石材由一条输送带送出，由卡车送往临近工地，作为混凝土的骨料，很快可以用到建设中。这种石材应该是石灰岩，是混凝土的重要组成部分，将来可部分重新应用于地底城的建设。开凿后剩下的石材，足以铺设三万多里马路，这个长度可以绕地球一圈半。这种方式的开凿使挖掘一座地底城变得容易，但新的问题是支撑现有的整座城市。

7.6 莫斯科的地下建设

芝加哥由于其又大又冷而闻名。但相比来说，有个地方更大更冷，那就是莫斯科。有着 1500 万人口的莫斯科都会区，加上 300 多万辆车足以造成最严重的堵塞，莫斯科失控般的扩张将把城市推向灾难边缘。

Mikhail Mikhailovich：现在真正的第一要务，就是解决市内运输系统崩溃的问题，以及无时无刻的道路堵塞。

为了解决这些问题，莫斯科只能向下发展。

Mikhail Mikhailovich：莫斯科是个文化古迹中心，如果我们想在不损毁古迹的前提下建造任何东西，就得向地下发展。

在当时，俄罗斯正兴起一股地下建筑风潮，有两项新计划正在推动，旨在防止城市受到发展的瓶颈干扰。其中之一就是托伏思凯亚购物中心，完工后将成为全欧洲最大的地下购物中心：有四层楼深，面积达 35 万平方尺。但进行这项建造方案，必须每天让行经附近道路的近 29 万 4 千辆车进行改道，施工人员必须紧靠着危险的马路工作。第二个建案，保罗列兹广场，是一个五层楼深的购物中心兼停车场，与地面上的车站相结合，但在这项计划实现之前，施工人员首先要解决该市柔软的土质问题。

Mikhail Mikhailovich：莫斯科的土质情况十分特殊，造成了施工困难，所以每个建造方案必须经过非常缜密的勘查。

松软的土质对市内特有的较低矮建筑不会造成问题，但并不适于建构地下建筑。如果要找到足够坚固的土壤层，就必须往下挖掘 90 多尺深进入较坚硬的黏土层。

Mikhail Mikhailovich：我唯一确定的是，在莫斯科建构地下城市并不是件困难的事，我们有各种必要的机具和足够的知识储备，因为我们从 20 世纪初就开始了建筑物的建造。

这座历史古城的瑰宝之一，就是位于数百尺深的地底莫斯科地铁。莫斯科拥有着全世界最繁忙的地铁系统之一。每天有 900 万人搭乘 4000 班车，行经 177 站，总长超过 190 里。随着需求的增加，地铁不断扩张，目前最新的一站是胜利公园，建设了 16 年于 2003 年启用。它拥有许多惊人的成就：全欧洲最长的电扶梯，耗时 4 分钟将乘客送往地下 328 尺的月台，也是当时全世界最深的地铁车站。

谈到保罗列兹广场的建设，施工人员对地下的挑战毫不陌生。直接往下挖个超大洞穴作为购物中心是不可行的。因为莫斯科松软的土壤很难固定。所以不是先凿洞，然后再在四周搭建混凝土墙，相反地，施工工人先把墙砌起来。一台开挖机的六对长爪挖进泥泞中，挖出二尺宽的洞后，马上灌入混凝土，四周形成墙面后，接着再从中间挖空。

工程缓慢又费工，经常挖挖停停进行清理工作。

一旦墙面固定成形，施工人员就能继续开挖，但即使墙面固定，危险也尚未解除。在城市中开挖，意味着要变更电缆线设施、排水系统和其他公共基础设施。最终，施工人员实现了保罗列兹购物中心中空的建设。就像在工地的另一端，施工人员已进入地下100尺深，用钢筋、混凝土和塑胶固定墙面。在仍有水渗入的条件下，向内坍塌的危险时刻存在。因此施工用直径五尺、长度约200尺的巨大钢柱来支撑墙面。巨型的钢柱仅仅是临时设施，一旦外墙整体固化，这些钢柱就会拆除。尽管保罗列兹广场离莫斯科河还有半里之遥，但由于水总能找到突破口，工程师时刻保持警觉。

视角回到芝加哥，这座城市就坐落在大湖旁，因此如何阻挡这些湖水进入地底城，将是一个大问题。湖水灌入将造成重大灾难。芝加哥地底城的每个楼层，必须加强固定。为了抢在人口暴增威胁使整个城市无法动弹之前，这座地底城必须加紧动工。

7.7 地下工程的通风技术

在地底城建设完成之前，还有一个严峻的问题需要解决。对任何广场而言，好的空气品质是首要条件。在施工过程中，施工人员可以采用附近伏尔坎广场既有的通风方式。垂直的通风井从通往广场的隧道抽入空气，直接往上送，确保粉尘和其他污染物不会停留太久，这个系统能够让连续工作十小时的数十名矿工能有舒适安全的工作环境。而考虑到将来24小时全天候生活在地底的数千居民，通风系统就变得复杂多了。

Charles Dowding：人们将在这个空间生活50到100年，而矿工在此施工不到十年。

在深层的地底，将没有敞开的窗户，没有微风可使新鲜流通。然而，幸运的是通风问题早在地底城建造之前便解决了。

在麻州波士顿，在堪称全美史上最昂贵的"大开挖工程"公共工程项目的背后，工程师每天面对在近四里长的隧道内数千辆汽车排放的数吨有毒气体。行进中的汽车可以推动隧道内空气流动，而堵车会造成空气停滞，使危险的废气堆积。于是麻州收费高速公路局安装了复杂的通风系统保证空气的流通。在隧道出口处附近，他们布置喷射风扇来循环空气。但由于有些隧道深达120尺，波士顿必须采取更先进的手段才能把空气送入地底。在波士顿市中心的某栋建筑物底，是隧道中心主干道的"肺"——实现这栋125尺高的建筑物吸入新鲜空气并将之送到地底的隧道。

Helmut Ernst：空气会被天窗吸进去，吸到地下的鼓风机室，经由一连串管路进入隧道。

在100尺地底的大型鼓风室，巨大的风扇将新鲜空气从通风井抽入。400马力的引擎每分钟可以运送33万3千立方尺的空气，形成的风速高达时速62里/小时。依靠古老的技术，这成为了一个强大的系统。

芝加哥地底城将会很好地利用波士顿的这项技术，只是地底的深度是波士顿的十倍。地底城的墙壁和地板将布满通风室，城市底部的风扇把新鲜空气从顶端的入风井抽入，空气将送入1300尺地底，在整座地底城里循环之后才会送到地面排出。不间断地为所有居民调控空气品质，不仅在于汽车交通，还需要更多的风扇以及完整的控制系统来充当整座城的眼睛和耳朵，能让芝加哥地底城所有居民有安全感。

7.10 对住在地下城市的关注

随着工程挑战一一突破，不仅提高了芝加哥地下生活的品质，还改善了地面交通，使市区街道不再拥挤不堪，交通不再堵塞，大众运输系统也得到喘息。人们通勤不再从市区外围进入，而是由下而上。

Charles Dowding：如果将大部分设施迁往地下，整座城市会更加美丽。将主要公共设施移往地下，可以腾出更多空间让我们高效利用地面。

就波士顿而言，完成大开挖工程之后，波士顿得到了重新的整合，一些地面土地得以回收作为公共用途，而不是一味地在上方架高架来扩张交通。曾经的钢结构建筑如今变成了公园。这些对地面市容的改善，成为支持地底城继续发展的有利力量。但在高楼大厦之下持续向下开凿地底城也是有风险的，因而地底城也必须寻找没有开发的地方实现拓展。

Charles Dowding：伊利诺伊州拥有整座湖，从湖边一路向外延伸，上面完全没有建筑物，或许这样的做法可以节省大量经费，并更快达成挖凿地底城的工程。

将新的地底空间网络与既有的地底城连接，规划人员可以在湖底建造另一座与之呼应的地底城。

Charles Dowding：要扭转世人对芝加哥的看法，可能要改变目前垂直式的美丽天际线，而开始思考地下天际线。

然而开发第二个地底城有一个很严峻的问题，地底城一期是在无人居住的情况下建造，而对于二期施工，原本对施工人员造成的威胁，现在将影响到地底城的数千永久居民。地底城在建设过程中无法像地面上那样可以看到全貌，所以建造时会遇到更多未知的困难。

最需解决的重要问题，就是埋藏在地底沉积已久的天然气。如果施工人员不小心碰到藏在地底的瓦斯管线，后果将不堪设想。

而（对于地底城）逃生计划的制定应该与现有摩天大楼的情况类似。

Charles Dowding：对于在大楼中设置多少部电梯，设置多少逃生楼梯等考虑，对于地底城来说也是同样适用的。

但在地底数千尺之下，这些思考必须"上下翻转"。由于遇难时电梯无法运作，楼梯将成为唯一的逃生路径。在地底城底层的人们必须爬上等同100层楼那样的高度逃生。对年幼、年长或残疾人来说是不可行的。紧急状况下需要市政当局英明的决断和作为，封锁灾害区域，保护城市其他区域免于有毒气体的危害。

前述讨论的技术挑战可以一一克服，但心理接受能力又是另一回事。问题是人们需要花多少时间接受地底的生活方式。在这座都市城市极速扩张和人口爆炸的困境下，这些疑虑可能会被现实需求覆盖。

终将有一天人们将不再只是在地底购物，或者在地底开车，而是会生活在那里。

Unit 8　香港海上机场

8.2　大型填海造陆工程

岛屿需要扎实的基底，于是他们必须清除黏土，挖出干净的基底。工程师找来全球最大的疏浚船队，把黏土全吸上来。疏浚船就像大型吸尘器，这些"吸尘器"的历史渊源来自现在的土耳其 800 年前的发明。当时的统治者想把一条河流上方的贫瘠土地化为茂盛的绿地。一位大胆的天才发明家 Alzazeri 接下了这个输水上山的挑战，他打造出全球第一座自动抽水泵。

Alhasani 教授致力于研究古代科技，他复制了古代的自动抽水泵系统，以脚踏车提供动力。脚踏车代表了当时的水车，即提供了要抽上山分配给各村镇的用水。活塞是这套机器的核心，像一支注射器，从管子吸入池塘的水，然后透过水箱从顶上的出水管排出。止回阀保证水只能向上流动。当搅动河水中的淤泥，泥巴连同水会一起被这个设备抽上来，于是疏浚的原理就来自于此。13 世纪的发明可以用到 20 世纪末香港机场的建设中来。正如水泵可以抽水，赤鱲角的疏浚大队可以利用此技术抽出海床的污泥，只不过后者的速度更快——连续两年稳定在一秒可抽出十吨的污泥。泥巴抽吸干净之后，工程师就有稳定的基底可以施工。到 1995 年，打算兴建香港新机场的岛屿填筑完成。

8.5　没有柱子支撑的轻型屋顶

剩下的任务是为新航站楼打造全球最大的封闭式空间。香港机场的设计者怀着雄心壮志，他们想打造的可不是普通的屋顶，而是全球最大的封闭式空间，同时还要这个空间轻盈、通风。

大空间需要大的屋顶，宽 700m、长 1km 以上，这可不是一个普通屋顶。机场的建筑师坚持要提出让人眼前一亮的设计——模拟中国南海起伏的波浪造型。但对于结构师来说，这将是一场噩梦：这么大的屋顶需要大量的支柱来支撑，但建筑师不想让笨重的结构及凌乱的柱子影响轻盈宽敞的航站楼的美观性。

航站楼应该是一个宽敞的空间，乘客在其中任意走动，没有任何阻碍。支撑的柱子之间应该保证有很大的间距。因此，究竟如何让少数几根极细的柱子支撑如此巨大的屋顶？答案就在于打造极为轻盈的屋顶，不过轻盈通常意味着屋顶不够坚固。

有趣的是，兼顾轻巧和坚固，这也向来是飞机设计师的基本工作。为了解决问题，设计师向"天上取经"，从第二次世界大战的一架轰炸机中找到灵感。在惠灵顿，一架英军的双引擎轰炸机，在第二次世界大战期间对德国的夜间进行空袭。机身的关键就在于其强度。

英国工程师 Barney Wallis 采取一种创新的方式打造轻巧而坚固的机身。机身的空洞比金属还多，但秘诀就在材料的排列方式，即重点不在于用了多少金属，而在于如何使用。一个试验向我们展示了如何使用少量的金属也可以发挥大用处。Martin Manning 是香港国际机场首席结构工程师，他作为试验顾问参与了这次试验。

首先在一根标准工字梁上加上荷载，荷载重 2.5 吨，相当于两辆家庭房车的重量。一根铁链吊着工字梁。随着铁链的松开，全部的荷载就由工字梁承担了。重达 130 千克的工字梁开始弯曲，勉强承受住了这 2.5 吨的荷载。

随后把荷载加到 3.5 吨，这时工字梁再也无法承受荷载。结果显示工字梁在 3.5 吨荷载时，已经完全破坏。为了提高它的承载力，通常可以把工字梁的尺寸加大，但就意味着工字梁的自重也加大了。谁也没想到，在可以实现让工字梁瘦身的同时，还增加它的强度。

联想惠灵顿的轰炸机，答案就藏在简单的几何学中。从 Barney Wallis 设计的惠灵顿轰炸机中获得启发，钢材重新组合成了一系列三角形的形状。组成后的格构梁比工字梁轻了近 20 千克。如果计算没错，它应该更加坚固。荷载会平均分布在格构梁上，所以不会折弯。接下来，这根格构梁将要检验其能否承受更大的荷载。

所有的砖块都被加载上，加起来共计 4.5 吨的荷载。当链条松开时，格构梁安然无恙。而在 3.5 吨荷载时，工字梁就已经完全破坏，估计它只有 2.75 吨左右的承载力。但格构梁在 4.5 吨荷载时仍纹丝不动，其承载力大概是工字梁的 2 倍。

钢材不同的排列方式对强度会造成重大差异。这根试验的格构梁采用非常坚固的三角形形状，就像惠灵顿的轰炸机。弧形的格构梁，同样也由三角形的形状组成，正是工程师所需要的用来建造香港机场航站楼屋顶。

每 36m 的跨度，都可以采用这巨大但轻巧的格构梁。格构梁能够实现自我支撑，不需要笨重的柱子，能让乘客在这个空间里自由活动。

8.7 风洞试验

但是这栋建筑物的外观尺寸带来另一个问题。台风，这些严重的热带暴风根据其摧毁力，在国际上有公认的等级划分。最高是第五级，风速可以增加到每小时 250km。这些台风在香港造成严重灾情。飓风也会影响侧面高耸平坦的大型建筑，就像这座香港机场一样。为了保护香港新机场不受破坏，工程师必须绞尽脑汁确保这栋建筑的牢固，但究竟要牢固到什么程度？

流动的空气可形成巨大的压力，为抵御这种荷载，侧墙需要加固。然而，强风对屋顶的威胁却是另外一种表现形式。飞机起飞靠的是升力，创造这种升力的秘诀之一是让空气快速流过表面。航站楼拥有类似这样的表面，在快速移动的空气中由于升力会造成十分严重的问题。

在风洞实验室里，将验证弧形的屋顶在台风级的强风下会发生什么。一个小的房子和一个外形类似香港机场的弧形屋顶作为试验对象。

主持人钻进了试验房子，这个风洞可以到达一级台风的风速，即每小时 130km。试验开始，起初屋顶没有异样，随着风速的加大，侧墙开始振动。时速刚过 55km/h，突然间屋顶就被掀开了。风力甚至把连接屋顶和房子的铁链扯断了。

主持人：当它过了那个临界点，你能感觉到那种升力。屋顶拼命要飞走，感觉就像翅膀。当时我能感觉到，在强风中有两股力量，因为刚开始风只朝房子的侧面直吹，想从侧面把房子吹垮，随着风力越来越强，终于达到临界点，使屋顶岌岌可危。一过临界点，强风就把屋顶掀开了。因此房子承受了两种不同的压力：一种是要摧毁侧墙墙壁，另一种是要掀开屋顶。

8.8　屋顶和墙壁的柔性连接

要把飘离的屋顶固定，有一个办法是把屋顶牢牢锁住，但这要求建筑物足够结实，这和建筑师的意图背道而驰。机场工程师想出了更好的办法，设计关键就在柔性连接，这要归功于1930年代一辆赛车首创的结合法。

这辆赛车叫银箭，一辆1930年代的梅赛德斯赛车。在它上面安装了梅赛德斯首创的一种革命性的新型悬吊系统。这个悬吊系统一开发出来，就迅速发展，从赛车应用领域迅速拓展到普通轿车上，至今仍在使用。

双A臂悬吊是一种三角形的强化钢材，只允许上下的位移，而阻止轮胎出现其他方向的移动。

一个小小的试验用来演示柔性和刚性之间的对比。有一则古老的寓言，说一棵坚固的橡树在暴风中被吹倒，小小的芦苇却撑了下来，因为它可以随风弯曲。以下这个演示试验就说明，柔性比刚性具有更好的防御能力。柔性连接就是香港国际机场采用的解决方案。

在试验中，首先将悬吊系统焊接到了一辆车上，轮胎和车子之间的连接不是柔性连接。试验的验证过程是将车辆吊起，然后摔到地上。结果在汽车重重摔到地上后，刚性连接让悬吊系统直接穿过引擎盖，车辆完全报废了。

相比于刚性连接，第二辆车有功能健全的悬吊系统。这一次，这辆车不但保持了柔性性能，在摔到地上后还可以继续行驶。

汽车的双A臂悬吊启发了机场设计者，把屋顶和墙壁采用柔性的结合方式连接，允许其在台风中有一定的位移。但不同于汽车的双A臂悬吊，机场工程师在设计时必须兼顾三个方向的位移。

屋顶在遭遇台风时，会像机翼似的浮起来，因此双A臂悬吊容许其上下移动，但猛烈吹向玻璃墙的风转化为一种横向运动，因此为了产生更大的柔性，它还具有一种滑动机制。采用铰式支座把双A臂悬吊连接到滑动杆。

航站楼总共设置1300副双A臂悬吊，让建筑物在台风的强风中弯曲不倒，全拜一辆古董赛车所赐。当飞机升空时，极少有旅客会知道这座机场是多了不起的成就。它的落成要归功于一座800年前的水泵，一支铜管乐队，一个冷战期间的窃听器，一架第二次世界大战的轰炸机以及一辆古董赛车。

Unit 9　防震工程

9.1　新知识来自新灾难

地震造成死亡和破坏。但纵观全世界，当工程师奋力对抗可怕的地震灾害时，建筑有防震方法，工业界也有防震方法，更重要的是，保护生命也有解决对策。仅仅通过地震模拟效果研究地震是不够的，地震工程师很清楚，想设计安全建筑就得研究真实发生的地震。

Steve Shekerlian：唯一有效的方法就是全尺模型，也就是通过真实的地震进行研究。

Prof. K. C. Tsai：地震工程是一个不断的学习过程，新知识来自新事件和新灾害。

Ron Egucchi：很不幸的是地震和大自然或许是研究地震的最佳测试环境，我们在实验室测试建筑和许多方法，但唯有在现实世界的大地震里，才能真正确定什么有效，什么无效。

1999年对地震工程师来说是忙碌的一年，第一次大地震发生在8月，土耳其人口密集区马尔马拉发生7.4级地震，45秒内，两万栋建筑倒塌，1万7千人丧命。在这种状况下做研究的有效性是其他工程不可比拟的，但也是个十分痛苦的任务。

Ron Egucchi：当我们到达土耳其时，就我个人而言，我非常不安，因为当地还有许多受损建筑，里面还有许多人受困，这让人很难过。

工程师的首要任务是进行灾区事后勘察，但进入灾害现场就像是进入战区。惨遭震害的社区已经面目全非，地震工程师有时无法辨认出自身所在的位置。在土耳其地震现场，Ron和同事利用全球定位系统和卫星地图才能正确找出需要勘察的坍塌位置。无需借助任何高科技设备，工程师都知道采用廉价建材的建筑几秒就能崩塌。许多地震表明不加钢筋的混凝土最不抗震。在高楼中，它会导致梁柱倒塌，造成楼板层层坍塌，这一现象称为"千层派塌陷"，也就是连续倒塌。

无论最终的破坏形式如何，令人不解的是，街道一边可能惨不忍睹，而另一边却毫发无伤。地震工程师明白，想了解其中原因必须从脚下的土地开始研究。地震的震波有P波（初波）和S波（次波），S波更猛烈，通常会毁坏建筑物，不同的土地类型会扩大或减小波的强度，土地越松软摇晃越大。就像声波和光波一样，震波也会反射和集中，某些社区可能因此遭受难以预测的严重损害。如果建筑横跨在断层上就会四分五裂；若建造于某种土壤上或许会陷入看似坚实的地下。这种危险而剧烈的过程称为"土壤液化"。土耳其公寓街区的底层楼就是遭遇此种变故。唯一拍摄到的土壤液化的实景，是在1963年日本新潟大地震，地震开始时，水由地底涌出。土壤液化在许多地震中造成建筑物倾斜和沉陷。美国最可怕的例子发生在旧金山的滨海区，当时是1989年洛马普利塔大地震，自然老师肖恩正要下班却发现停车场发生土壤液化波动，而他深陷其中。

Ken Finn：那是约5点钟，我在车里准备回家，车子上下摇动，好像有人在保险杠上跳舞，我意识到是地震了，地面晃得好厉害，根本无法行走，走动时必须很小心，停车场上下起伏约有3尺，范围大概有6尺见方，地面起起落落如同驻波，整个停车场都是这样。

Ken脚下的土壤松软饱含水分，水位接近地面。地震开始时，松软的土壤粒子与水分离，然后较重的粒子下沉，水位上升至地面，如同柔软的果冻一样。在易液化区，地震工程师将桩基打入地底来避免其影响。

9.4 加利福尼亚地震预防措施

土耳其马尔马拉或中国台湾大地震的资料对开发新的灾害模拟软件有很大帮助，一般的笔记本电脑就能告诉屋主地震对建筑的影响。

Steve是加州地震工程师，他能利用这项科技预测地震对全美任何地方的影响。例如一栋典型的加州建筑。

Steve Shekerlian：今天我们利用科技可以判断建筑可能受损的程度。有很多电脑软件都能实现这项功能，你可以输入建筑的地址以及建筑类型和年代。以一栋典型建筑为例，电脑能指出它和灾害的相关性。对软件的操作过程很简单：输入资料，找出建筑的位置，输入各个结构属性，然后储存资料，接着输入地震灾害类型，载入500年一遇的断层资料。之后，电脑根据灾害资料做数百次模拟分析，你就能预估确定这类型建筑在地震时的可能损失，以及不同水准下的受损程度和金钱损失。

但知道可能的损失和采取预防措施是两回事，现实中很多人住在高危险区并不采取防震措施，使家庭处在地震所带来的危险中。在地震中最惨重的倒塌或许是高楼的街区或高架桥，但事实上，对于加州来说，受损最多的建筑类型却是低楼层房屋，即单门独院的木建造屋。在洛杉矶，为鼓励居民对地震做充足的准备，市政府首次试行住宅防震分级制度，防震做的越好，等级越高，若是防震级别太低，房子将来可能无法出售，而相反，级数越高，则可以更多地可以保障家人安全。

9.6 高科技工程的尖端抗震技术

从西雅图到圣地亚哥，从台北到东京，很多高科技工业区都位于地震带，这些工业需要最有效的防震措施。发生在中国台湾的1999年的大地震，严重影响了号称"台湾硅谷"的新竹科学园区。新竹园区供应了全球三分之一的晶片，一旦停产将造成全球性影响。当时地震来袭时，电力中断，区域陷入一片黑暗。对晶圆厂而言，停电意味着停产。在停产两天，损失达到数百万美元之后，政府做出了空前的决定，将岛上仅存的电力输送到新竹。但电力并非是晶圆厂唯一的担忧，厂内有许多需要保护的昂贵物品。

John Dizon是地震工程师，负责其中一个大型工厂的地震事宜。

John Dizon：这（保护设备）比维护生命安全还复杂，许多设备是全年无休地运作，停产对运营受阻影响很大。公司希望地震后能马上复工。

对地震工程师而言，晶圆厂是很大的挑战，管线输送的气体有毒而且具有高腐蚀性，必须装设与地震仪相连的紧急关闭阀。一旦有小小的地震便停止输送。虽然有斜撑可以保护建筑，但当室内的东西很重要时，它们自身也必须采取一定的防护措施。

John Dizon：这些设备比建筑更昂贵，占制造厂资产总额很大的比例。

在生产晶片的无尘室里，一小粒灰尘也会造成严重的破坏，因此需要更高标准的防震。

John Dizon：无尘室有许多固定装置来保护昂贵的设备，例如防止在地震时滑落或倾覆的固定装置等。

如果说晶圆厂的室内物品非常昂贵，那么世界级研究实验室的室内物品或许就是无价之宝了。

伯克利大学位于海瓦德断层之上，最近的地震调查使大家都警觉起来，因为最有价值的新生命研究计划很容易在地震中受损。

Prof. Mary Comerio：伯克利校园研究惊奇地发现所有的赞助研究都集中在一起，约有百分之五十的研究集中在7栋建筑里。

如同台湾的晶圆厂一样，伯克利大学的研究也是24小时运转的，研究设备非常昂贵。如果这里发生地震，知识的损失是无价的。

Prof. Mary Comerio：我们无法承受研究大楼被关闭，我们不能把研究中止两年。若是装有样本的冰箱倾倒，癌症解药也许会随之而去。

这对未来的地震工程师来说是崭新的挑战，这种防震要求远远超过从前，必须确保大楼在地震中毫发无伤。

Prof. Mary Comerio：我们的目标是找出可持续运作的抗震方法。

解决方法就在尚未测试过的尖端地震工程技术中。

地震工程师的终极目标是建造抗震建筑。要想建筑在任何地震中都屹立不倒，现在已经有很多惊人的防震措施可以采用，当然还有一些妙招在逐渐成形。基底隔震是分离建筑与地面的技术，地面在大地震中会摇晃，但建筑不会跟着摇晃。在当时，洛杉矶市政厅是采用基底隔震的最高建筑。由于是旧建筑，补强加固是一项很大的挑战。

Richard Puckowski：一般安装隔震系统，是从地面向上建起，但洛杉矶市政厅建造于1928年，安装隔震器有很大的操作限制，建筑不能移动，原有建筑尽量不能受损。

当地震来袭时，市政厅要能在任何水平方向上平移4尺。为了实现这个诉求，在建筑四周开挖了一条4尺宽的壕沟。

Richard Puckowski：我们要实现建筑在地面发生位移时可以有足够的占据空间。想象一下（在地震中）建筑垂直不动而地面发生位移，要想建筑留在原地，它必须占有足够的空间，因此市政厅周围需要一条4尺宽的壕沟。

下一步要分离建筑与原有地基，工程师必须在原有地下室下面开挖，在原有柱子下方放置千斤顶以支撑建筑。下一步578个隔震器安装在原有地基和新的底层之间。这时，可以移除千斤顶，让建筑坐落在隔震器上。每个隔震器由橡胶与钢材组成间层，外面包上防火外壳，地震来袭时，隔震器底部会和地面一起摇动，隔震器的变形使得了建筑几乎纹丝不动。

Richard Puckowski：底层地下室的高度约4至5尺，安置有隔震器的平面隔离了整栋建筑。当地震来临时，这个平面以下会和地面一起摇动，地震结束时地基回到原有状态，建筑压向隔震器底部，最后达到平衡而静止，这一切发生在几微秒之内。

在建筑防震上日益重要的另一个尖端技术是阻尼器。阻尼技术约在20年前开始应用在军事工程上，使洲际导弹地下发射窖在受到近距离攻击时仍能运作。地震工程师利用阻尼器吸振，避免建筑在地震时摇晃。阻尼器通俗讲就是隔震器。一种阻尼器通过液体从小洞一端流到另一端，达到吸震的目的。由于挤压力推动液体流动，液体会变热，能量由动能转换成热能。阻尼器安装在建筑里能吸收共振减少摇晃。如果阻尼器的液体分子能按照不同的需求而改变特性，工程师就能使部分建筑变成刚性，其余仍保持柔性，让建筑对地震作出最佳反应，这就是智慧型材料。

Andrew Whittaker：应用在地震工程的智慧型材料，它的特性能即时改变。

工程师们已发现有材料在外力作用下能在几毫秒内改变特性。

Andrew Whittaker：以这两个注射管内的液体为例，前后抽动注射管，液体会在两个注射管内来回移动。当在两个管中间放上一块磁铁时，很难推动液体，液体的属性大大改变了。这在未来，给地震工程师提供绝佳的机会去进一步改善对结构反应的控制。

控制的关键在于强大电脑结合智慧型材料，或许真正的智慧型大楼即将到来。当然电脑也让工程师思考其他可行的解决办法。

地震工程经历了快速的发展，通过仔细收集的地震资料和先进的电脑技术，工程师离抗震建筑的目标越来越近，未来有许多惊人的可能性。

9.7　未来地震工程的科技畅想

George Lee 博士是未来地震工程研究的权威专家，主要研究主动控制。他专门钻研生物科技，将工程原理应用在人体上。他希望建筑能向人类学习，他尤其赞叹人类的平衡能力。Lee 博士每次搭电车上班时就注意到这一点：人类靠肌肉系统来平衡身体，大脑和肌肉共同合作来控制身体平衡。

Prof. George Lee：我们能看、能听、能感觉，感官接收讯息传到脑部，控制中枢告诉各个肌肉做不同程度的收缩，一些肌肉完全放松，一些肌肉收缩，弯曲膝盖表示你要增加缓冲，这样才能减轻冲击。

打太极的人对屈膝习以为常，其实这是人体工程的伟大之处。

Prof. George Lee：我在搭乘地铁时，会感到左右摇晃。这种惯性力，是重量乘以水平方向的加速度。这个力会移动我的身体，为了平衡水平方向的力量，我必须调整三个变量：重量、缓冲性和刚性。地铁若是突然开动，我会前后晃动，我会在这个方向调节平衡。

Lee 博士正在进行一项创举，将这一系统应用在建筑上。正如神经系统把信息传到脑部，建筑周围的传感器把信息传给电脑。电脑将与液压和配重系统连接，这相当于人体的肌腱和肌肉。

Prof. George Lee：当地震发生时，传感器测量加速度和速度，把信息传给电脑，然后大脑会决定该行动了，例如，按下 1 号控制器意味着多一点缓冲性，2 号制动器是多一点刚性，3 号制动器是部分刚性，诸如此类。建筑在应对地面摇动时就能产生最佳平衡。

这项技术称为主动控制，能使建筑做出实时反应。在日本和中国台湾省的一些建筑，已经安装了这项新科技的第一代。调谐质量阻尼器属于主动控制中的一种，感应器在地震时会把信息传给中央电脑，电脑会启动液压系统，在轨道上移动大金属块的配重。工程师相信地震来袭时，这可以抵消共振。

在当时，台北正在盖世界上最高的大楼，这里是地震最活跃的城市之一。这个大楼采用传统抗震技术与尖端技术相结合。摩天大楼采用最大的超级柱子支撑，并以密集的钢筋作为骨架。但由于地上有 88 层楼高，大楼里也会安装调谐质量阻尼器。具体效果如何要等到下次地震的检验。

同时 Lee 博士也希望摩天大楼能够学会如何"摔倒"，正如我们跌倒时会保护头部，他希望建筑也能有本能反应，在知道自己倒塌时保护最有价值的资产。

Prof. George Lee：建筑有时不可避免地会倒塌。如果你想保护某个储存有重要仪器的楼层，你可以进行个性化程序控制，例如让某些东西稳固、某些放松、其余的随它倒塌。

未来的路还很长，先进科技目前还十分昂贵，只是少数人的专利。大多数还是负担不起，因此世界各地的地震工程师都在寻找接地气的方法。

Ron Egucchi：当问大多数地震工程师有何希望时，答案常常是希望那些科技和设计，能应用于所有的结构中。

还有些工程师希望下一代不必生活在建筑倒塌的恐惧中。

Prof. K. C. Tsai：地震本身不会使人丧命，是人所建造的结构使人丧命。我们有时候必须预先进行规划，这样我们的孩子才不必忍受大地震之苦。

但对所有的地震工程师而言，他们都肩负着相同的使命。

Steve Shekerlian：我们应该保护大众安全，当大家进入到我们设计的建筑时，他们应该感到安全，知道建筑设计正确，不会倒塌。随着科技的日新月异，我们会不断提高，抗震建筑的目标终将达成。

Unit 11　FIDIC 合同在海外项目管理的应用

11.1　菲迪克简介

FIDIC（中译"菲迪克"）是国际咨询工程师联合会的法文首字母缩写。

菲迪克是由欧洲三个国家的咨询工程师协会于 1913 年成立的。组建联合会的目的是共同促进成员协会的职业利益，以及向其成员协会会员传播有益信息。

今天，菲迪克（FIDIC）已有来自于全球各地 60 多个国家的成员协会，代表着世界上大多数私人执业的咨询工程师。

菲迪克举办各类研讨会、会议及其他活动，以促进其目标：维护高的道德和职业标准；交流观点和信息；讨论成员协会和国际金融机构代表共同关心的问题；以及发展中国家工程咨询业的发展。

菲迪克的出版物包括：各类会议和研讨会的文件，为咨询工程师、项目业主和国际开发机构提供的信息，资格预审标准格式，合同文件，以及客户与工程咨询单位协议书。这些资料可以从设在瑞士的菲迪克秘书处得到。

11.2　菲迪克合同的标准格式及其适用范围

《施工合同条件》

推荐用于由业主或其代表工程师设计的建筑或工程项目。这种合同的通常情况是，由承包商按照业主提供的设计进行工程施工。但该工程可以包含由承包商设计的土木、机械、电气和（或）构筑物的某些部分。

《生产设备和设计-施工合同条件》

推荐用于电气和（或）机械设备供货和建筑或工程的设计与施工。这种合同的通常情况是，由承包商按照业主要求，设计和提供生产设备和（或）其他工程；可以包括土木、机械、电气和（或）构筑物的任何组合。

《设计采购施工（EPC）/交钥匙工程合同条件》

可适用于以交钥匙方式提供工厂或类似设施的加工或动力设备、基础设施项目或其他类型开发项目，这种方式（a）项目的最终价格和要求的工期具有更大程度的确定性，（b）由承包商承担项目的设计和实施的全部职责，业主介入很少。交钥匙工程的通常情况是，由承包商进行全部设计、采购和施工（EPC），提供一个配备完善的设施，（"转动钥匙"时）即可运行。

《简明合同格式》

推荐用于资本金额较小的建筑或工程项目。根据工程的类型和具体情况，这种格式也可用于较大资本金额的合同，特别是适用于简单或重复性的工程或工期较短的工程。这种合同的通常情况是，由承包商按照业主或其代表（如果有）提供的设计进行工程施工，但这种格式也可适用于包括或全部是由承包商设计的土木、机械、电气和（或）构筑物的合同。

这些合同格式是推荐在国际招标中通用的。在某些司法管辖范围，特别是用于国内合同的条件，可能需要做些修改。菲迪克认为，正式的、权威性的文本应为英文版。

在编写《施工合同条件》的过程中意识到，虽然有许多条款可以通用，但是有些条款必须考虑特定合同的有关情况做出必要的改变。可以用于多数（但非全部）合同的条款已包括进通用条件中，以便纳入每项合同。

通用条件和专用条件共同组成管理合同各方权利和义务的合同条件。对每个具体的合同都需要编制其专用条件，要考虑那些提到专用条件的通用条件条款。

11.3 菲迪克合同术语的准确释义

"合同"系指合同协议书、中标函、投标函、本条件、规范、图纸、资料表以及合同协议书或中标函中列出的其他文件（如果有）。

"中标函"系指业主签署的正式接受投标函的信件，包括其所附的含有双方间签署的协议的任何备忘录。如无此类中标函，则"中标函"一词系指合同协议书，发出或收到中标函的日期系指签署合同协议书的日期。

"投标函"系指由承包商填写的名为投标函的文件，包括其签署的向业主的工程报价。

"规范"系指包含在合同中名为规范的文件，以及按照合同对规范所作的任何补充和修改。此类文件规定对工程的要求。

"图纸"系指包含在合同中的工程图纸，以及由业主（或其代表）按照合同发出的任何补充和修改的图纸。

"资料表"系指合同中名为资料表的文件，由承包商填写并随投标函一起提交。此类文件可包括工程量表、数据、表册、费率和（或）价格表。

"投标书"系指投标函和合同中包括的由承包商随投标函一起提交的所有其他文件。

"投标书附录"系指填写的名为投标书附录的文件，附在投标函后作为其一部分。

"工程量表"和"计日工作计划表"系指在资料表中具有这一名称的文件（如果有）。

"当事方（一方）"根据上下文需要，指业主或承包商。

"业主"系指在投标书附录中称为业主的当事人，及其财产所有权的合法继承人。

"承包商"系指在业主接受的投标函中称为承包商的当事人，及其财产所有权的合法继承人。

"分包商"系指为完成部分工程，在合同中指名为分包商，或其后被任命为分包商的任何人员；以及这些人员各自财产所有权的合法继承人。

"分项工程"系指在投标书附录中确定为分项工程的工程组成部分。

"承包商文件"系指由承包商根据合同提交的所有计算书、计算机程序和其他软件、图纸、手册、模型和其他技术性文件（如果有）。

履约担保

承包商应对严格履约（自费）取得履约担保，保证金额和币种应符合投标书附录中的规定。承包商应在收到中标函后28天内向业主提交履约担保，并向工程师送一份副本。履约担保应由业主批准的国家（或其他司法管辖区域）内的实体提供，并采用专用条件所附格式，或业主批准的其他格式。

变更

在颁发工程接收证书前的任何时间，工程师可通过发布指示或要求承包商提交建议书的方式，提出变更。

承包商应遵守并执行每项变更，除非承包商立即向工程师发出通知，说明（附详细根据）承包商难以取得变更所需的货物。工程师接到此类通知后应取消、确认或改变原指示。

每项变更可包括：

（a）合同中包括的任何工作内容的数量的改变（但此类改变不一定构成变更）；

（b）任何工作内容的质量或其他特性的改变；

（c）任何部分工程的标高、位置和（或）尺寸的改变；

（d）任何工作的删减，但要交他人实施的工作除外；

（e）永久工程所需的任何附加工作、生产设备、材料或服务，包括任何有关的竣工试验、钻孔和其他试验和勘探工作；

（f）实施工程的顺序或时间安排的改变。

除非并直到工程师指示或批准了变更，承包商不得对永久工程作任何改变和（或）修改。

11.4 文件管理

文件管理是管理行为的载体，也是践行合约精神的重要体现，优质的文件管理能够使工程管理工作更加流畅和高效。

文件优先次序

构成本合同的文件要认为是互作说明的。为了解释的目的，文件的优先次序如下：

（a）合同协议书（如果有）

（b）中标函

（c）投标函

（d）专用条件

（e）本通用条件

（f）规范

（g）图纸

（h）资料表和构成合同组成部分的其他文件

如文件中发现有歧义或不一致，工程师应发出必要的澄清或指示。

文件的照管和提供

规范和图纸应由业主保存和照管。除非合同中另有规定，应给承包商一式两份合同文本和后续图纸，承包商可以自费复制或要求提供更多份数。

除非并直到被业主接收为止，每份承包商文件都应由承包商保存和管理。除合同中另有规定外，承包商应向工程师提供一式六套承包商文件。

承包商应在现场保存一份合同、规范中指名的出版物、承包商文件（如果有）、图纸、变更，以及根据合同发出的其他往来文书。业主人员有权在所有合理的时间使用所有这些文件。

如果一方发现为实施工程准备的文件中有技术性错误或缺陷，应立即将该错误或缺陷通知另一方。

Appendix 2: Common Terms in Civil Engineering
附录 2：土木工程常用术语

一般术语

1. civil engineering　土木工程
2. building and civil engineering structures　工程结构
3. design of building and civil engineering structures　工程结构设计
4. building engineering　房屋建筑工程
5. highway engineering　公路工程
6. railway engineering　铁路工程
7. port (harbour) and waterway engineering　港口与航道工程
8. hydraulic engineering　水利工程
9. hydraulic and hydroelectric engineering　水利发电工程（水电工程）
10. construction work　建筑物（构筑物）
11. structure　结构
12. foundation　基础
13. foundation soil; ground　地基
14. timber structure　木结构
15. masonry structure　砌体结构
16. steel structure　钢结构
17. concrete structure　混凝土（砼）结构
18. special engineering structure　特种工程结构
19. building　房屋建筑
20. industrial building　工业建筑
21. civil building; civil architecture　民用建筑
22. highway　公路
23. highway network　公路网
24. freeway　高速公路
25. arterial highway　干线公路
26. feeder highway　支线公路
27. railway; railroad　铁路（铁道）
28. standard gauge railway　标准轨距铁路
29. broad gauge railway　宽轨距铁路
30. narrow gauge railway　窄轨距铁路
31. railway terminal　铁路枢纽
32. railway station　铁路车站

33. port; harbour 港口
34. marine structure 港口水工建筑物
35. navigation structure; navigation construction 通航（过船）建筑物
36. light house 灯塔
37. water conservancy 水利
38. multipurpose hydraulic project; key water-control project; hydro-junction 水利枢纽
39. reservoir 水库
40. hydraulic structure; marine structure; maritime construction 水工建筑物
41. water retaining structure; retaining works 挡水建筑物
42. intake structure 进水（取水）建筑物
43. outlet structure; outlet works; sluice works 泄水建筑物
44. conveyance structure 输水建筑物
45. regulating structure; training structure rectification structure 整治建筑物
46. hydro-electric station; hydropower station 水电站
47. pump station 水泵站（抽水站、扬水站、提水站）
48. raftpass facility log pass facility 过木建筑物（过木设施）
49. fishpass facility 过鱼建筑物（过鱼设施）
50. safety device 安全设施

房屋建筑结构术语

1. composite structure 组合结构
2. hybrid structure 混合结构
3. slab-column system 板柱结构
4. frame structure 框架结构
5. arch structure 拱结构
6. folded-plate structure 折板结构
7. shell structure 壳体结构
8. space truss structure 风架结构
9. cable-suspended structure 悬索结构
10. pneumatic structure 充气结构
11. shear wall structure 剪力墙（结构墙）结构
12. frame-shear wall structure 框架-剪力墙结构
13. tube structure 筒体结构
14. suspended structure 悬挂结构
15. high-rise structure 高耸结构

公路路线和铁路线路术语

1. highway 公路，道路，大路，路线
2. highway alignment 公路线形

3. horizontal alignment　平面线形
4. vertical alignment　纵面线形
5. route selection　公路选线
6. route location　公路定线
7. horizontal curve　平面线
8. vertical curve　竖曲线
9. grade change point　变坡点
10. route intersection　路线交叉：两条或两条以上公路的交会
11. permanent way　铁路线路
12. railway location　铁路选线
13. location　铁路定线
14. main line　正线
15. sidings　站线
16. minimum radius of curve　最小曲线半径
17. grade section　坡段
18. maximum grade　最大坡度
19. grade crossing　平面交叉
20. grade separation　立体交叉

桥、涵洞和隧道术语

1. bridge　桥
2. simple supported girder bridge　简支梁桥
3. continuous girder bridge　连续梁桥
4. cantilever girder bridge　悬臂梁桥
5. cable stayed bridge　斜拉（斜张）桥
6. suspension bridge　悬索（吊）桥
7. trussed bridge　桁架桥
8. frame bridge　框架桥
9. rigid frame bridge　刚构（刚架）桥
10. arch bridge　拱桥
11. submersible bridge　漫水桥
12. pontoon bridge　浮桥
13. right bridge　正交桥
14. skew bridge　斜交桥
15. grade separated bridge; overpass bridge　跨线（立交）桥
16. viaduct　高架桥
17. main span　正（主）桥
18. approach span　引桥
19. curved bridge　弯桥

20. ramp bridge　坡桥
21. combined bridge；highway and railway transit bridge　公路铁路两用桥
22. movable bridge　开合桥
23. single-track bridge　单线桥
24. double-track bridge　双线桥
25. bridge superstructure　桥跨结构（上部结构）
26. bridge floor system　桥面系统
27. bridge bearing；bridge support　桥支座
28. bridge substructure　桥下部结构
29. bridge tower　索塔（桥塔）
30. abutment　桥台
31. pier　桥墩
32. culvert　涵洞
33. tunnel　隧道（洞）
34. tunnel portal　隧道洞口（洞门）
35. tunnel surrounding rock　隧道（洞）围岩
36. tunnel lining　隧道（洞）衬砌
37. viscous damper　黏滞阻尼器

水工建筑物术语

1. dam　坝
2. dam axis　坝轴线
3. gravity dam　重力坝
4. arch dam　拱坝
5. buttress dam　支墩坝
6. earth-rock dam；embankment dam　土石坝
7. concrete dam　混凝土坝
8. rubber dam；flexible dam；fabric dam　橡胶坝
9. spur dike；groin　丁坝
10. training dike　顺坝
11. spillway　溢洪道
12. weir　堰（溢流堰）
13. coffer dam　围堰
14. hydraulic tunnel　水工隧洞
15. deep water intake　深式进水口
16. dam type hydropower station　堤坝式水电站
17. diversion conduit type hydropower station　引水（引水道）式水电站
18. tidal power station　潮汐电站
19. pumped storage power station　抽水蓄能电站

20. powerhouse of hydropower station　水电站厂房
21. forebay　前池，前舱
22. pressure conduit　压力管道
23. surge chamber　调压室
24. tailrace　尾水渠
25. navigation lock　船闸
26. ship lift; ship elevator　升船机
27. sluice; barrage　水闸
28. channel　渠道
29. aqueduct; bridged flume　渡槽
30. chute　陡坡
31. drop　跌水
32. gallery system　坝内廊道系统
33. energy dissipating and anti-scour facility　消能防冲设施
34. seepage control facility　防渗设施
35. drainage facility　排水设施
36. reverse filter　反滤设施（倒滤设施）
37. turbine-pump station　水轮泵站
38. ram station　水锤泵站
39. under dam culvert　坝下埋管
40. silting basin　沉消池
41. dike; levee　堤
42. breakwater; mole　防波堤
43. wharf; quay　码头
44. sloped wharf　斜坡码头
45. dolphin wharf　墩式码头
46. gravity quay-wall　重力式码头
47. sheet-pile quay-wall　板桩码头
48. open pier on piles; high-pile wharf　高桩码头
49. floating pier; pontoon wharf　浮码头
50. dock　船坞
51. ship-building berth　船台
52. slipway　滑道

结构构件和部件术语

1. member　构件
2. component; assembly parts　结构中由若干构件组成的组合件
3. section　截面
4. beam; girder　梁

5. arch   拱
6. slab; plate   板
7. shell   壳
8. column   柱
9. wall   墙
10. truss   桁架
11. frame   框架
12. bent frame   排架
13. rigid frame   刚架（刚构）
14. simply supported beam   简支梁
15. cantilever beam   悬臂梁
16. beam fixed at both ends   两端固定梁
17. continuous beam   连续梁
18. superposed beam   叠合梁
19. pile   桩
20. sheet pile   板桩
21. pavement   路面
22. carriageway   行车道
23. speed-change lane   变速车道
24. sidewalk   人行道
25. lane separator   分隔带
26. bicycle path   自行车道
27. road shoulder   公路路肩
28. subgrade side ditch   路基边沟
29. catch ditch; intercepting channel   截水沟（天沟）
30. drainage ditch   排水沟
31. slope protection; revetment   护坡
32. retaining wall   挡土墙
33. railway track   铁路轨道
34. rail   钢轨
35. sleeper   轨枕
36. track skeleton   轨排
37. bed   道床
38. ballast   道碴
39. turnout   道岔
40. railway shunting hump   铁路调车驼峰
41. continuous welded rail   无缝线路
42. rail fastening   钢轨扣件
43. guard rail   护轮轨

44. railway shoulder　铁路路肩
45. wharf shoulder　码头胸墙
46. relieving slab　卸荷板
47. berthing member　靠船构件
48. mooring post bollard　系船柱
49. mooring ring　系船环
50. sluice chamber　闸室
51. sluice gate；lock gate　闸门
52. sluice pier　闸墩
53. apron　护坦
54. apron extension　海漫
55. stilling basin　消能池（消力池）
56. roller bucket　辊斗式消能器
57. apron；impervious blanket　防渗铺盖
58. impervious curtain；cut-off　防渗帷幕
59. sealing；seal；waterstop　止水带，止水条
60. connection　连接
61. joint　节点
62. expansion and contraction joint　伸缩缝
63. settlement joint　沉降缝
64. aseismic joint　防震缝
65. construction joint　施工缝

地基和基础术语

1. spread foundation　扩展（扩大）基础
2. rigid foundation　刚性基础
3. single footing　独立基础
4. combined footing　联合基础
5. strip foundation　条形基础
6. shell foundation　壳体基础
7. box foundation　箱形基础
8. raft foundation　筏形基础
9. pile foundation　桩基础
10. open caisson foundation　沉井基础
11. cylinder pile foundation；cylinder caisson foundation　管柱基础
12. caisson foundation　沉箱基础
13. subgrade of highway（railway）　路基
14. bed；bedding　基床

结构可靠性和设计方法术语

1. reliability　可靠性
2. safety　安全性
3. serviceability　适用性
4. durability　耐久性
5. basic variable　基本变量
6. design reference period　设计基准期
7. probability of survival　可靠概率
8. probability of failure　失效概率
9. reliability index　可靠指标
10. calibration　校准法
11. deterministic method　定值设计法
12. probabilistic method　概率设计法
13. permissible（allowable）stresses method　容许应力设计法
14. ultimate strength method　破坏强度设计法
15. limit states method　极限状态设计法
16. limit states　极限状态
17. limit state equation　极限状态方程
18. ultimate limit states　承载能力极限状态
19. serviceability limit states　正常使用极限状态
20. partial safety factor　分项系数
21. design situation　设计状况
22. persistent situation　持久状况
23. transient situation　短暂状况
24. accidental situation　偶然状况

结构上的作用、作用代表值和作用效应术语

1. action　作用
2. load　荷载
3. force per unit length　线分布力
4. force per unit area　面分布力
5. force per unit volume　体分布力
6. moment of force　力矩
7. permanent action　永久作用
8. variable action　可变作用
9. accidental action　偶然作用
10. fixed action　固定作用
11. free action　自由（可动）作用

12. static action　静态作用
13. dynamic action　动态作用
14. repeated action; cyclic action　多次重复作用
15. low frequency cyclic action　低周反复作用
16. self weight　自重
17. site load　施工荷载
18. earth pressure　土压力
19. temperature action　温度作用
20. earthquake action　地震作用
21. explosion action　爆炸作用
22. wind load　风荷载
23. wind vibration　风振
24. snow load　雪荷载
25. crane load　吊车荷载
26. floor live load; roof live load　楼面、屋面活荷载
27. load on bridge　桥（桥梁）荷载
28. dead load on bridge　桥（桥梁）恒荷载
29. live load on bridge　桥（桥梁）活荷载
30. standard highway vehicle load　公路车辆荷载标准
31. standard railway live load specified by the People's Republic of China　中国铁路标准活载
32. ship load　船舶荷载
33. crane and vehicle load　起重运输机械荷载
34. ship impact force　船舶撞击力
35. ship breasting force　船舶挤靠力
36. mooring force　船舶系缆力
37. water pressure　水压力
38. buoyancy　浮力
39. uplift pressure　扬压力
40. wave pressure; wave force　浪压力（波浪力）
41. ice pressure　冰压力
42. silt pressure　泥沙压力
43. frost heave force; frost heave pressure　冻胀力
44. representative value of an action　作用代表值
45. characteristic value of an action　作用标准值
46. quasi-permanent value of an action　作用准永久值
47. combination value of actions　作用组合值
48. partial safety factor for action　作用分项系数
49. design value of an action　作用设计值

50. coefficient for combination value of actions　作用组合值系数
51. effects of actions　作用效应
52. coefficient of effects of actions　作用效应系数
53. normal force　轴向力
54. shear force　剪力
55. bending moment　弯矩
56. bimoment　双弯矩
57. torque　扭矩
58. stress　应力
59. normal stress　正应力
60. shear stress; tangential stress　剪应力
61. principal stress　主应力
62. prestress　预应力
63. displacement　位移
64. deflection　挠度
65. deformation　变形
66. elastic deformation　弹性变形
67. plastic deformation　塑性变形
68. imposed deformation　外加变形
69. restrained deformation　约束变形
70. strain　应变
71. linear strain　线应变
72. shear strain; tangential strain　剪应变
73. principal strain　主应变
74. combination for action effects　作用效应组合
75. fundamental combination for action effects　作用效应基本组合
76. accidental combination for action effects　作用效应偶然组合
77. combination for short-term action effects　短期效应组合
78. combination for long-term action effects　长期效应组合
79. limiting design value　设计限值

材料性能、构件承载能力和材料性能代表值术语

1. resistance　抗力
2. strength　强度
3. compressive strength　抗压强度
4. tensile strength　抗拉强度
5. shear strength　抗剪强度
6. flexural strength　抗弯强度
7. yield strength　屈服强度

8. fatigue strength　疲劳强度
9. ultimate strain　极限应变
10. modulus of elasticity　弹性模量
11. shear modulus　剪变模量
12. modulus of deformation　变形模量
13. Poisson ratio　泊松比
14. bearing capacity　承载能力
15. compressive capacity　受压承载能力
16. tensile capacity　受拉承载能力
17. shear capacity　受剪承载能力
18. flexural capacity　受弯承载能力
19. torsional capacity　受扭承载能力
20. fatigue capacity　疲劳承载能力
21. stiffness；rigidity　刚度
22. crack resistance　抗裂度
23. ultimate deformation　极限变形
25. spatial behavior　空间工作性能
26. brittle failure　脆性破坏
27. ductile failure　延性破坏
28. partial safety factor for resistance　抗力分项系数
29. characteristic value of a property of a material　材料性能标准值
30. partial safety factor for property of material　材料性能分项系数
31. design value of a property of a material　材料性能设计值
32. nominal value of geometric parameter　几何参数标准值

几何参数和常用量程术语

1. height of section；depth of section　截面高度
2. breadth of section　截面宽度
3. thickness of section　截面厚度
4. diameter of section　截面直径
5. perimeter of section　截面周长
6. area of section　截面面积
7. first moment of area　截面面积矩
8. second moment of area；moment of inertia　截面惯性矩
9. polar second moment of area；polar moment of inertia　截面极惯性矩
10. section modulus　截面模量（抵抗矩）
11. radius of gyration　截面回转半径
12. eccentricity　偏心矩
13. relative eccentricity　偏心率

14. span 跨度
15. rise 矢高
16. slenderness ratio 长细比
17. longitudinal gradient 纵坡
18. superelevation 超高
19. sight distance 视距
20. width of subgrade 路面宽度
21. width of subgrade 路基宽度
22. clearance of highway 公路建筑限界
23. gauge 轨距
24. railroad clearance 铁路建筑限界
25. clearance under bridge 桥下净空
26. construction height of bridge 桥建筑高度
27. clearance above bridge floor 桥建筑限界
28. clearance of tunnel 隧道建筑限界
29. berth 泊位
30. additional depth; residual depth 富余水深
31. wave characteristics; wave parameters 波浪要素
32. tide level 潮位
33. water level 水位
34. design water level 设计水位
35. dam height 坝高
36. dam length 坝长
37. free board 安全超高（富余高度）
38. dead water level 水库死水位
39. normal (pool) level 水库设计（正常）蓄水位
40. design flood level 水库设计洪水位
41. exceptional flood level 水库校长核洪水位
42. dead storage 水库死（垫底）库容
43. usable storage 水库兴利（有效、调节）库容
44. total reservoir storage 水库总库容

工程结构设计常用的物理学、数理统计

1. coefficient of friction 摩擦系数
2. mass density 质量密度
3. force (weight) density 重力密度
4. moment of momentum 动量矩
5. dynamic moment of inertia 转动惯量
6. dynamic effect factor 动作用系数

7. vibration   振动
8. acceleration   加速度
9. frequency   频率
10. natural frequency   自振（固有）频率
11. period   周期
12. natural period of vibration   自振周期
13. periodic vibration   周期振动
14. amplitude of vibration   振幅
15. degree of freedom   自由度
16. damp   阻尼
17. forced vibration   强迫振动
18. mode of vibration   振型
19. resonance   共振
20. statistical parameter   统计参数
21. mean value   平均值
22. mean square deviation   方差
23. standard deviation   标准差
24. coefficient of mean value   均值系数
25. coefficient of variation   变异系数
26. probability distribution   概率分布
27. fraction   分数
28. significance level   显著性水平
29. hydro-static pressure   静水压强
30. hydro-dynamic pressure   动水压强
31. total hydro-static pressure   静水总压力
32. pressure gradient   压力梯度
33. pressure head   压力水头
34. level head   位置水头
35. stream field   流场
36. streamline   流线
37. velocity of flow   流速
38. velocity head of flow   流速水头
39. total head   总水头
40. head loss   水头损失
41. discharge cross section   过水断面
42. wetted perimeter   湿周
43. hydraulic radius   水力半径
44. discharge; flow rate   流量
45. average velocity   平均流速

46. coefficient of roughness　糙率（粗糙系数）
47. hydraulic slope; energy gradient　水力坡度（水力比降）
48. Reynolds Number (Re)　雷诺数
49. Froude Number (Fr)　弗汝德数
50. water hammer　水锤（水击）
51. hydraulic jump　水跃
52. seepage flow　渗流
53. coefficient of compressibility　压缩系数
54. cohesion　内聚力（黏聚力）
55. coefficient of consolidation　固结系数
56. relative density　相对密度
57. compactness　密实度
58. modulus of compressibility　压缩模量
59. porosity　孔隙比
60. porosity　孔隙率（度）
61. liquidity index　液性指数
62. plasticity　塑性指数
63. degree of saturation　渗透系数
64. degree of saturation　饱和度
65. degree of consolidation　固结度
66. pore water pressure　孔隙水压力
67. water content　含水量
68. liquid limit　液限
69. plastic limit　塑限
70. angle of repose　休止角
71. angle of external friction　外摩擦角
72. angle of internal friction　内摩擦角
73. earthquake　地震
74. earthquake focus　震源
75. earthquake centre　震中
76. epicenter distance　震中距
77. earthquake magnitude　地震震级
78. earthquake intensity　地震烈度
79. earthquake zone　地震区
80. earthquake response spectrum　反应谱
81. static method　静力法
82. equivalent base shear method　底部剪力法（拟静力法）
83. time-history method　时程分析法
84. mode analysis method　振型分解法

85. earthquake dynamic water pressure　地震动水压力
86. earthquake dynamic earth pressure　地震动土压力
87. liquefaction of saturated soil　砂土液化

Appendix 3: Network Resource
附录3：网络资源

1. EI	http://www.ei.org
2. SCI	http://www.webofknowledge.com
3. CNKI	http://www.cnki.net/
4. SDL Free Translation	http://www.freetranslation.com
5. Babel Fish	http://www.babelfish.com
6. Word Reference	http://www.wordreference.com
7. Bing	https://www.bing.com/translator/
8. Systranet	https://goo.gl/FHbGUE
9. Spell Check Plus	http://spellcheckplus.com/
10. Grammerly	https://www.grammarly.com/
11. Online Correction	http://www.onlinecorrection.com/

References
参 考 文 献

[1] National Geographic. Mega Structures - Skyscraper. 2004. Web.
[2] National Geographic. Mega Structures - Mega Bridges：China. 2004. Web.
[3] National Geographic. Mega Structures - Autobahn. 2004. Web.
[4] Discovery Channel. Man Made Marvels - World's fastest Railway. 2008. Web.
[5] National Geographic. Mega Structures-Inside the Channel Tunnel. 2004. Web.
[6] Discovery Channel. Mega Engineering - Chicago Underground. 2009. Web.
[7] National Geographic. Engineering Connections - Hong Kong's Ocean Airport. 2009. Web.
[8] Discovery Channel US - Engineering Against Earthquakes. 2006. Web.
[9] 牛津英语联想词典[M]. 北京：商务印书馆，2003.
[10] 牛津实用英汉双解词典[M]. 北京：外语教学与研究出版社，2007.
[11] 英汉土木建筑大词典[M]. 北京：中国建筑工业出版社，1999.
[12] 牛津高阶英汉双解词典[M]. 北京：商务印书馆，2004
[13] 刘润清，何福胜，张敬源. 当代研究生英语[M]. 北京：外语教学与研究出版社，2000
[14] 鲁正. 土木工程专业英语[M]. 北京：机械工业出版社，2018.
[15] 宿晓萍. 土木工程专业英语[M]. 北京：北京大学出版社，2017.
[16] 俞家欢. 土木工程专业英语[M]. 北京：清华大学出版社，2017.
[17] 崔春义. 土木工程专业英语[M]. 北京：中国建筑工业出版社，2015.
[18] 苏小卒. 土木工程专业英语[M]. 上海：同济大学出版社，2015.
[19] 李锦辉，陈锐. 土木工程专业英语[M]. 上海：同济大学出版社，2012.
[20] 雷自学. 土木工程专业英语[M]. 北京：知识产权出版社，2010.
[21] 李亚东. 土木工程专业英语[M]. 成都：西南交通大学出版社，2005.
[22] 董祥. 土木工程专业英语[M]. 南京：东南大学出版社，2011.
[23] 郭向荣. 土木工程专业英语[M]. 北京：中国铁道出版社，2011.
[24] 段兵延. 土木工程专业英语[M]. 武汉：武汉理工大学出版社，2010.
[25] 胡庚申. 论文写作与国际发表[M]. 北京：外语教学与研究出版社，2014.
[26] 刘振聪，修月祯. 英语学术论文写作[M]. 北京：中国人民大学出版社，2013.
[27] Assaf SA, Al-Hejji S. Causes of delay in large construction projects[J]. Project Manage, 2006, 24(4)：349-357.
[28] Al-Momani A. Construction delay：a quantitative analysis[J]. Project Manage, 2000, 20：51-59.
[29] Frimpong Y, Oluwoye J, Crawford L. Causes of delay and costoverruns in construction of groundwater projects in a developingcountries；Ghana as a case study[J]. Project Manage, 2003, 21：321-326.
[30] Odeh AM, Battaineh HT. Causes of construction delay：traditionalcontracts[J]. Project Manage, 2002, 20：67-73.
[31] Manavazhia MR, Adhikarib DK. Material and equipment procurement delays in highway projects in Nepal [J]. Project Manage, 2002, 20：627-632.
[32] Leite, Fernanda, Akcamete, Asli, Akinci, Burcu, Atasoy, Guzide, Kiziltas, Semiha. Analysis of Modeling Effort and Impact of Different Levels of Detail in Building Information Models[J]. Automation in Con-

struction, 2011.

[33] Smith, Deke. An Introduction to Building Information Modeling (BIM)[J]. Journal of Building Information Modeling, 2007.

[34] Assaf SA, Al-Hejji S. Causes of delay in large construction projects[J]. Project Manage, 2006, 24(4): 349-357.

[35] Ajanlekoko JO. Controlling cost in the construction industry[J]. LagosQS Digest, 1987, 1(1): 8-12.

[36] Odeyinka HA, Yusif A. The causes and effects of construction delayson completion cost of housing project in Nigeria[J]. Financial ManageProperty Construction, 1997, 2(3): 31-44.

[37] Ogunlana SO, Promkuntong K. Construction delays in a fast growingeconomy: comparing Thailand with other economies[J]. ProjectManage, 1996, 14(1): 37-45.

[38] Al-Momani A. Construction delay: a quantitative analysis[J]. Project Manage, 2000, 20: 51-59.

[39] Frimpong Y, Oluwoye J, Crawford L. Causes of delay and costoverruns in construction of groundwater projects in a developingcountries: Ghana as a case study[J]. Project Manage, 2003, 21: 321-326.

[40] Chan DWM, Kumaraswamy MM. A comparative study of causes oftime overruns in Hong Kong construction projects[J]. ProjectManage, 1997, 15(1): 55-63.

[41] Akinsola AO. Neural network model for predicting building projects' contingency[C]. Proceedings of Association of Researchers in Construction Management, ARCOM 96, Sheffield HallamUniversity, England, 11-13 September 1996: 507-516.

[42] Alkass S, Mazerolle M, Harris F. Construction delay analysistechniques[J]. Construction Manage Econ, 1996, 14(5): 375-394.

[43] Odeh AM, Battaineh HT. Causes of construction delay: traditionalcontracts[J]. Project Manage, 2002, 20: 67-73.

[44] Manavazhia MR, Adhikarib DK. Material and equipment procurement delays in highway projects in Nepal [J]. Project Manage, 2002, 20: 627-632.

[45] http://bbs.zhulong.com/106010-group-918/detail30245085/

[46] http://www.bimcn.org/BIMal/201608297454.html

[47] http://www.autodesk.com/case-studies/shanghai-tower-construcfion-development

[48] http://en.wikipedia.org/wiki/Building-information-modeling

Acknowledgements
致　　谢

　　本书的出版是编者以及研究生们共同努力的成果。除本书所列参考文献外，还广泛参考和借鉴了国内外大量的文献资料，在此对其相关作者一起表示感谢。需要特别说明的是，第 1 章是在鲁修红教授、余泽川博士、胡琴博士的共同协作下完成的；研究生郑挚（第 2 章）、谢少军（第 3 章和第 7 章）、蔡小明（第 4 章和第 5 章）、罗丽英（第 6 章）、李婉玲（第 8 章）、徐乐（第 9 章）、姜日鑫（第 10 章）、吴伟波（第 11 章）完成了英文的初步整理。本科生夏延完成了本书的图片处理，吴昊学和刘智杰完成了相关的视频处理。在此向他们辛勤付出表示衷心的感谢。

　　特别感谢武汉大学徐礼华教授、浙江大学徐世烺教授、湖南大学易伟建教授、湖北工业大学校长刘德富教授、肖本林教授的勉励和支持，他们都为本书内容的不断完善提出了宝贵的修改建议。

　　感谢肖衡林教授、贺行洋教授、马强教授、胡其志教授、李扬副教授、陈娜博士等的鼓励和支持。

　　感谢教材分社副社长王跃编审（博士）给予的专业指导和真诚帮助，感谢赵莉编辑严谨的专业指导。

编者　夏冬桃
2019 年 7 月